International Workshop on Flavour Changing and Conserving Processes (FCCP2022)

AA003647

EPJ Web of Conferences Volume 289 (2023)

Anacapri, Italy
22-24 September 2022

Editors:

G. D'Ambrosio
M. Iacovacci
S. Mastroianni
M. Passera
G. Venanzoni

ISBN: 978-1-7138-8045-5

Printed from e-media with permission by:

Curran Associates, Inc.
57 Morehouse Lane
Red Hook, NY 12571

Some format issues inherent in the e-media version may also appear in this print version.

For additional information, please contact EDP Sciences – Web of Conferences
at the address below.

EDP Sciences – Web of Conferences
17, Avenue du Hoggar
Parc d'Activité de Courtabœuf
BP 112
F-91944 Les Ulis Cedex A
France

Phone: +33 (0) 1 69 18 75 75
Fax: +33 (0) 1 69 28 84 91

contact-edps@webofconferences.org

Additional copies of this publication are available from:

Curran Associates, Inc.
57 Morehouse Lane
Red Hook, NY 12571 USA
Phone: 845-758-0400
Fax: 845-758-2633
Email: curran@proceedings.com
Web: www.proceedings.com

TABLE OF CONTENTS

PREFACE

The fourth edition of the International Workshop on "Flavour changing and conserving processes" 2022 (FCCP2022) took place on Sep 22-24, 2022 in Anacapri, Capri Island, Italy. The Workshop was organized by the Physics Department of the University of Naples "Federico II" and by INFN (Istituto Nazionale di Fisica Nucleare) – Sezione di Napoli. The involved scientific community has been that of experiments like muon g−2, Mu2e, MEG, Mu3e, COMET, but also that of the K and B meson physics. More than thirty talks were presented covering the theory status as well as the latest results of experiments in various Laboratories of the world. Major attention was drawn on the muon g-2 first result from Fermilab. We wish to thank everybody who contributed to the success of the Workshop.
We look forward to the next edition waiting for new results from many sides, especially on the muon g-2 front.

Giancarlo D'Ambrosio
Michele Iacovacci
Massimo Passera
Graziano Venanzoni
Stefano Mastroianni

Editors
FCCP2022

Statement of Peer review

In submitting conference proceedings to EPJ Web of Conferences, I certify to the Publisher that I adhere to the **Policy on Publishing Integrity** of the journal in order to safeguard good scientific practice in publishing.

1. All articles have been subjected to peer review administered by the proceedings editors.
2. Reviews have been conducted by expert referees, who have been requested to provide unbiased and constructive comments aimed, whenever possible, at improving the work.
3. Proceedings editors have taken all reasonable steps to ensure the quality of the materials they publish and their decision to accept or reject a paper for publication has been based only on the merits of the work and the relevance to the journal.

Title, date and place of the conference FCCP2022, 22-24 September 2022, Anacapri, Capri (IT)

Proceedings editor(s): G. D'Ambrosio, M. Iacovacci, S. Mastroianni, M. Passera, G. Venanzoni

October, 23 2023

Date and editor's signature

17, avenue du Hoggar - PA de Courtabœuf – BP 112 - 91944 Les Ulis Cedex A (France)
Tél. : 33 (0)1 69 18 75 75 - Fax : 33(0)1 69 07 45 17 – www.edpsciences.org

Status of the Muon $g-2$ experiment

Peter Winter[1,*]

[1] *Argonne National Laboratory, Lemont, IL 60439, USA*

Abstract. The Muon $g-2$ Experiment at Fermi National Accelerator Laboratory was designed to measure the anomalous magnetic moment of the muon, a_μ, with a precision of 140 parts-per-billion; a four-fold improvement over the former BNL measurement. The Fermilab experiment was motivated by the about 3.5 standard deviation between the experiment and the Standard Model calculation of a_μ which could be a hint of new physics. The experiment at Fermilab relies on the well-established storage ring technique using magic momentum muons but employs new detector systems and a higher rate of muons per injection cycle to achieve the significant improvement in precision. A first result from the Run-1 data taking period has achieved an uncertainty of 0.46 parts-per-million and confirmed the BNL discrepancy, further increasing the tension with the Standard Model to $4.2\,\sigma$. The experimental technique, key aspects of the measurement, and the data analysis of Run-1 will be summarized.

1 Introduction

The magnetic moment $\vec{\mu}$ of a charged elementary particle is linked to its spin $\vec{S}$ via the gyro-magnetic ratio g. For the structure-less, spin-$\frac{1}{2}$ electron, the Dirac equation predicted $g_e \equiv 2$ [1] while virtual particle loops give rise to small additional contributions known as the anomalous magnetic moment, $a_\mu = \frac{1}{2}(g_\mu - 2)$. The dominant contribution, the so-called Schwinger term, is of size $\alpha/2\pi$ [2] and arises from the coupling of a virtual photon to the muon.

The understanding of the anomalous magnetic moments of electrons and muons have played an essential role in the formulation of our current understanding of fundamental particles and their interactions, the so-called Standard Model (SM) of Particle Physics. The measurement of the electron's anomalous magnetic moment [3] offers one of the most precise determinations of the fine structure constant, α_{QED} or in conjunction with other measurements of α [4, 5] the most precise test of QED. The muon anomalous magnetic moment, a_μ, is more sensitive to contributions from heavier particles by a factor of $(m_\mu^2/m_e^2) \approx 42000$ compared to the electron. As many extensions to the SM predict heavy new particles, a correct prediction of a_μ is an important test for any new theory.

On the experimental side, there has been a long-standing history of continual improvements in the precision of a_μ. The latest milestone prior to the E989 Muon g-2 experiment at Fermilab was reached by the Brookhaven E821 experiment [6]. The about $3.5\,\sigma$ deviation between the experiment and the SM calculation was a possible hint of new physics and motivated the new Muon $g-2$ experiment at Fermilab, which was designed with a precision goal of 140 part-per-billion (ppb) [7]. Assuming that the central values of both SM and experiment remained the same this improved precision would lead to $5\,\sigma$ signal of new physics. The first result [8–11] from the new experiment is based on the analysis of the Run-1 data. The result confirmed the BNL measurement with similar precision. The difference of the experimental world average of $(116,592,061 \pm 41) \times 10^{-11}$ (350 ppb) with the current SM [12–32] is $(251 \pm 59) \times 10^{-11}$ or $4.2\,\sigma$ as shown in Fig. 1.

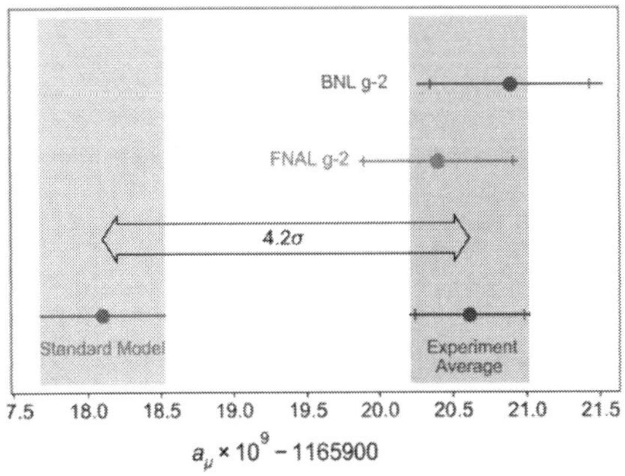

Figure 1. Run-1 result for a_μ from the Fermilab experiment together with the former BNL result [6], the new weighted experimental average, and the SM value [12].

The SM calculation for the anomalous magnetic moment, a_μ, includes electromagnetic, weak, and strong corrections The largest theoretical uncertainty stems from the first-order hadronic vacuum polarization (HVP), which is evaluated from low energy e^+e^- data via a dispersion relation. Significant improvement in the precision in re-

*e-mail: winterp@anl.gov

cent years came from new, precise data from the CMD and SND experiments at Novosibirsk, KLOE at Frascati, BaBar at SLAC, and Belle at KEK. A recent result from CMD-3 [33] for the two pion channel shows a significant discrepancy to other experimental measurements in this low-energy regime and will require further investigations.

In past years, lattice-QCD has also made significant progress but results using this approach were not as precise as the dispersive approach and hence not yet used in the current SM value for the HVP [12]. However, a lattice-QCD result by the BMW collaboration with sub-percent precision [34] was released in 2021 and showed tension with the HVP value obtained from the data-driven approach using the dispersion relation. Recently, the lattice-QCD community focused on the so-called intermediate window energy range which reduces the dominant systematic uncertainties entering the full HVP determination. Results for the determination of a_μ in the intermediate window [34–41] show confirmation of the BMW value and further effort is needed to understand the discrepancy between lattice-QCD and data-driven determinations of the HVP. More details on the theoretical work for a_μ are presented in other contributions to these proceedings.

2 The Fermilab Muon $g-2$ experiment

The Fermilab Muon $g-2$ experiment is based on the storage ring technique established at CERN [42–44] and BNL [6]. To reach the fourfold improvement in precision on a_μ, many of the detection systems used at BNL were replaced or significantly upgraded.

2.1 Muon production, injection, and storage

The muons are produced at Fermilab via proton collisions in the former anti-proton production target. Positive pions produced in the target are momentum selected and decay into muons in the transport beam line to the experimental hall. Due to parity violation of the weak force, the resulting, highly polarized muon beam is injected into the superconducting 1.45 T storage ring shown in Fig. 2, which was relocated from Brookhaven.

The muons enter the storage ring through an inflector magnet [45]. The inflector is a non-ferrous, double cosine theta, superconducting magnet that's placed inside the main magnet yoke to cancel the storage B-field. The resulting field-free channel offers the muons a path to enter the storage area of the main magnet. Since they are displaced radially outwards by 77 mm from the ideal orbit, the muons would be lost during their first turn. A set of 3 new, fast kicker magnets [46] were built to deflect the muons after a quarter turn onto the ideal orbit. The system uses a Blumlein design to deliver the fast kicking pulse with a pulse length of less than 149 ns, the revolution time of the muons in the storage ring. Pulsed, electric quadrupoles [47] are employed to contain the muons vertically.

The relevant observable for the anomalous precession frequency, ω_a, is the rate of change of $\hat{\beta} \cdot \vec{S}$, where $\hat{\beta}$ and

Figure 2. An image of the storage ring during the Run-1 data taking period. Credit: Reidar Hahn, Fermilab.

$\vec{S}$ are the directions of the muon's momentum and spin, respectively. Since the angle between the muon's spin precession, $\vec{\omega}_s$, and its cyclotron precession, $\vec{\omega}_c$, is small, the rate change of $\hat{\beta} \cdot \vec{S}$ can be approximated by $\vec{\omega}_a$, which in the presence of the magnetic, $\vec{B}$, and electric field, $\vec{E}$, is:

$$\vec{\omega}_a \equiv \vec{\omega}_s - \vec{\omega}_c = -\frac{q}{m}\left[a_\mu\vec{B} - \left(a_\mu - \frac{1}{\gamma^2-1}\right)\frac{\vec{\beta}\times\vec{E}}{c} - a_\mu\left(\frac{\gamma}{\gamma+1}(\vec{\beta}\cdot\vec{B})\vec{\beta}\right)\right] \quad (1)$$

For horizontally circulating muons in a vertical magnetic field, $\vec{\beta} \cdot \vec{B} = 0$, rendering the term in the second line zero. This condition is approximately met in the Muon g-2 experiment. To overcome the term from the motional magnetic field induced by the electric field, $\vec{\beta} \times \vec{E}/c$, the CERN-III, BNL, and Fermilab experiments use muons at the "magic" momentum ($\gamma = 29.3$, $p = 3.096\,\mathrm{GeV}/c$), where $a_\mu - 1/(\gamma^2 - 1) = 0$. A future $g-2$/EDM experiment at J-PARC [48] takes a different approach using weak magnetic focusing, e.g. $\vec{E} = 0$. With these conditions met, Eq. (1) simplifies Eq. (1) to

$$\omega_a = \frac{e}{m}a_\mu|\vec{B}|, \quad (2)$$

showing that the determination of a_μ requires precise measurements of the anomalous spin precession frequency of the muons, ω_a, and the magnetic field, $\vec{B}$. Corrections due to vertical motion of the muons (pitch correction) and the off-magic momentum distribution (E-field correction) will be discussed in Sec. 3.

2.2 Muon anomalous precession and magnetic field measurements

The muons mainly decay via $\mu \to e\nu\bar{\nu}$ with a time-dilated lifetime of $\tau \approx 64\,\mu\mathrm{s}$. Due to the parity violating weak decay, the positrons' direction carries an energy-dependent asymmetry A(E) with respect to the muon's spin. The experiment is equipped with 24 calorimeter stations placed radially inwards with respect to the muon storage region. Each calorimeter consists of 54 PbF_2 Cherenkov crystals readout with silicon photomultipliers (SiPM) [49–51]. The signals from the SiPMs are digitized in custom-made 800 mega-samples per second digitizers. The calorimeters determine the the energy and time of the decay positrons and the segmentation suppressed systematic effects from pile-up. The measured positron rate in the calorimeters is an exponential modulated by ω_a. A fit to the data allows for the extraction of ω_a as detailed in Sec. 3. The gain of each calorimeter crystal is monitored with a sophisticated laser calibration system [52]. The major gain systematics are the infill gain change of order 10 µs and double pulse gain change of order 10 ns.

The precise measurement of the averaged magnetic field as seen by the stored muon ensemble is achieved by using nuclear magnetic resonance (NMR) of protons in water, a technique pioneered by Bloch [53] and Purcell [54]. The averaged azimuthal field distribution is mapped with an in-vacuum field camera [55], the so-called trolley. It is equipped with 17 NMR probes and provides detailed maps of the magnetic field during periods without stored muons every 3-5 days. 378 fixed NMR probes continuously monitor the field above and below the muon storage aperture during the trolley maps and muon injection. An additional probe with a cylindrical water sample [56] provides the absolute calibration of the magnetic field in terms of the shielded proton Larmor frequency, ω'_p. A novel ^{3}He NMR calibration probe [57] provides a second calibration.

Information on the muon beam distribution is mainly accessed through two straw tracker stations [58] located inside the vacuum chamber 90 degrees apart in front of two calorimeter stations. While the trackers measure the decay positron trajectories, the information can be used to infer on the muon beam properties. Further information on the beam dynamics of the stored muons can be obtained from insertable fiber harp detectors located at 180 and 270 degrees from the injection point and the Injected Beam Monitoring System with one station each before and after the inflector magnet and an insertable third station in the storage region closely behind the injection point. The information obtained from these systems provides important input for determining several of the beam dynamics related corrections needed to obtain a_μ. A short summary of these corrections is provided in Sec. 3.3.

3 Run-1 data analysis summary

3.1 Analysis overview

Equation (2) is the basis for the determination of a_μ. It can be rewritten in terms of a ratio of two frequencies [11] and external, well-known constants as:

$$a_\mu = \frac{\omega_a}{\tilde{\omega}'_p(T_r)} \frac{\mu'_p(T_r)}{\mu_e(H)} \frac{\mu_e(H)}{\mu_e} \frac{m_\mu}{m_e} \frac{g_e}{2}. \qquad (3)$$

Here, $\tilde{\omega}'_p(T_r)$ represents the magnetic field experienced by the stored muons in terms of the Larmor frequency of shielded protons in a spherical water sample at temperature T_r = 34.7 °C. The tilde indicates that this quantity is weighted by the actual muon distribution observed in the experiment. The ratio $\mu_e(H)/\mu'_p(T)$ of the magnetic moments of an electron bound in hydrogen to that of a proton shielded in a spherical water sample is measured to 10.5 ppb [59]. The bound-state QED corrections that determine the magnetic moment ratio of the electron bound in hydrogen versus a free electron, $\mu_e(H)/\mu_e$, are considered essentially exact [60], and the electron magnetic moment μ_e is known to 0.3 ppb [60]. The ratio of the mass of the muon and the mass of the electron, m_μ/m_e, is known to 22 ppb from the measurement of the hyperfine splitting of muonium [61] and bound-state QED [60]. Finally, the g factor of the electron, g_e is known to 0.13 ppt [3].

Equation (3) requires the measurement of the ratio $\omega_a/\tilde{\omega}'_p(T_r)$ to determine a_μ. Several terms enter into the determination of this ratio which can be conceptually written as [8]:

$$\frac{\omega_a}{\tilde{\omega}'_p(T_r)} \approx \frac{f_{\rm clock} \cdot \omega_a^m \cdot (1 + C_e + C_p + C_{ml} + C_{pa})}{f_{\rm calib} \cdot \langle \omega_p(x, y, \phi) \times M(x, y, \phi) \rangle \cdot (1 + B_k + B_q)}. \qquad (4)$$

Here, $f_{\rm clock}$ represents a factor to compensate for the secret, slight detuning of the main experimental clock that was done to guarantee a fully blind analysis and ω_a^m is the measured anomalous spin precession frequency as outlined more in Sec. 3.2. The beam dynamics related corrections C_e, C_p, C_{ml}, and C_{pa} are further detailed in Sec. 3.3. The denominator comprises the factor $f_{\rm calib}$, which accounts for the calibration of the trolley probes with the calibration probe. The term $\langle \omega_p(x, y, \phi) \times M(x, y, \phi) \rangle$ denotes the weighting of the field maps, $\omega_p(x, y, \phi)$, measured with the trolley and fixed probes with the beam distribution, $M(x, y, \phi)$, obtained from the straw detectors. Details about this analysis are provided in Sec. 3.4. The final corrections B_k and B_q stem from fast magnetic field transients from the kicker and quadrupole systems, respectively. More details will be given in Sec. 3.5 for B_q whereas B_k is described in detail in another contribution to this conference. Table 1 provides a summary for the Run-1 result for the corrections (C and B terms in Eq. (4)) and the uncertainties of the quantities in Eq. (4) and the external constants from Eq. (3).

3.2 Measuring ω_a^m

The measurement of ω_a is centered around the measured spin precession frequency, ω_a^m, as can be seen in the nominator of Eq. (4). The determination of ω_a^m succeeds through the measurement of the energy and time of the decay positrons in the 24 calorimeter stations. One of these calorimeter stations is shown in the left panel Fig. 3.

Due to the parity violating weak decay, high energy positrons are emitted preferentially along the direction of the muon spin, hence offering a means to observe the spin precession. For the analysis, the time distribution of decay positrons above an energy threshold of $E \sim 1.7$ GeV is histogrammed as shown in the right panel of Fig. 3.

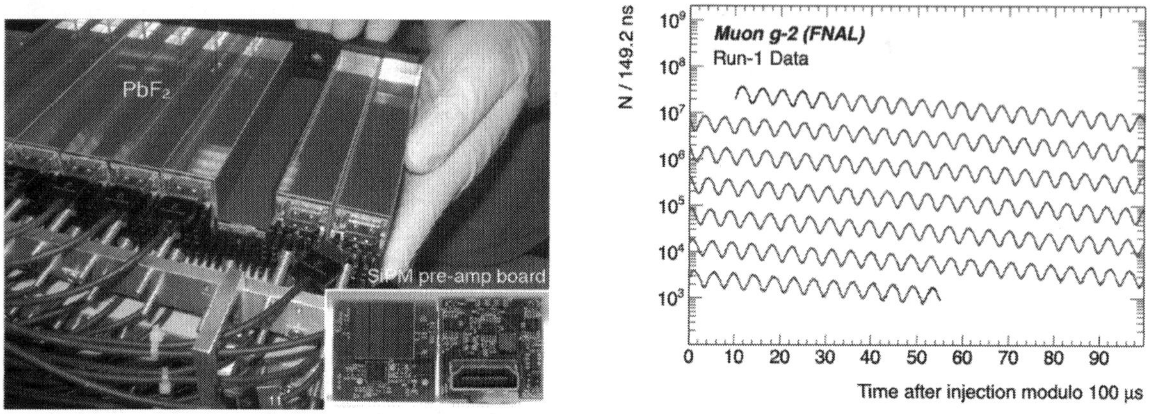

Figure 3. Left panel: Assembly of one of the 24 calorimeter stations each consisting of 54 PbF_2 crystals. The small insets show the silicon photomultipliers pre-amplifier board that are mounted on each crystal. Right panel: A typical wiggle plot - i.e., time distribution of detected positrons - for the Run-1 dataset.

Table 1. Summary of the Run-1 corrections and uncertainties for the quantities in Eq. (4) and external constants from Eq. (3).

Quantity	Correction (ppb)	Uncert. (ppb)
ω_a^m (stat.)	-	434
ω_a^m (syst.)	-	56
C_e	489	53
C_p	180	13
C_{ml}	-11	5
C_{pa}	-158	75
$f_{calib}\langle\omega_p(x,y,\phi) \times M(x,y,\phi)\rangle$	-	56
B_k	-27	37
B_q	-17	92
$\mu_p'(T_r)/\mu_e(H)$	-	10
m_μ/m_e	-	22
$g_e/2$	-	0
Totals	544	462

This so-called wiggle plot is the basis for the determination of ω_a^m and its main features can be described by the muon lifetime τ and the spin precession, ω_a^m, via $N(t) = N_0(t)e^{t/\tau}[1 - A\cos\omega_a^m t + \phi]$. The full analysis requires a more complicated fit function to account for various beam dynamics effects [10].

3.3 Beam dynamics corrections

A short summary of the four beam dynamics corrections, C_e, C_p, C_{ml}, and C_{pa} listed in the nominator of Eq. (4) is given here. Both the electric field correction C_e and the pitch correction C_p are related to the fact that Eq. (2) is only valid for ideal conditions of a horizontally circulating, magic-momentum muon in a vertical magnetic field. Given the momentum distribution and emittance of the muon beam, the full analysis must also take into account the additional terms in Eq. (1) as they do not fully vanish.

The muon momentum spread of $\Delta p/p \sim 0.15\,\%$ leads to not all muons being exactly on the magic momentum such that the $\vec{\beta} \times \vec{E}$ term has to be taken into account. The derivation of the correction is data driven and based on the equilibrium radius $\langle x_e^2\rangle$ of the muons. This quantity can be measured from the debunching of the initial, injected muon bunch. Two methods, one based on Fourier transform [62] and a direct fit of the debunching signal [43, 63] determined $\langle x_e^2\rangle$ with good agreement as shown in the left panel of Fig. 4. From the measured $\langle x_e^2\rangle$ distribution, an electric field correction $C_e = 489 \pm 53$ ppb was derived for the Run-1 result [9].

The pitch correction originate from the $\vec{\beta} \cdot \vec{B}$ term in Eq. (1) as the stored muons have a vertical angular distribution. Its determination is related to the vertical muon distribution [64, 65] which can be determined from the vertical decay position of electrons measured with the straw trackers, as shown in the middle panel of Fig. 4. The more involved determination includes acceptance corrections and other elements as detailed in [9] and led to a Run-1 correction of $C_p = 180 \pm 13$ ppb.

The number of stored muons during each injection cycle reduces not only through their decay into positrons but also due to some muons losing energy in elements like the beam shaping collimators or other materials and no longer being stored. These lost muons can carry different average phases from early to late in each muon fill. This would cause a change in the ω_a frequency over time. As these lost muons are no longer stored, they travel radially inward and being minimum ionizing particles, they leave small amounts of energy in neighboring calorimeters. Hence, a time coincidence of events in two or three neighboring calorimeters provides a way to measure the rate of lost muons and led to a correction of $C_{ml} = -11 \pm 5$ ppb.

The final correction is related to a movement of the muon beam during the storage time. The main origin for this beam movement in Run-1 was due to broken resistors in the electrostatic quadrupoles which led to a change in the quad high-voltage during muon storage and hence a change in their focusing strength. The determination of the so-called phase-acceptance correction included the use

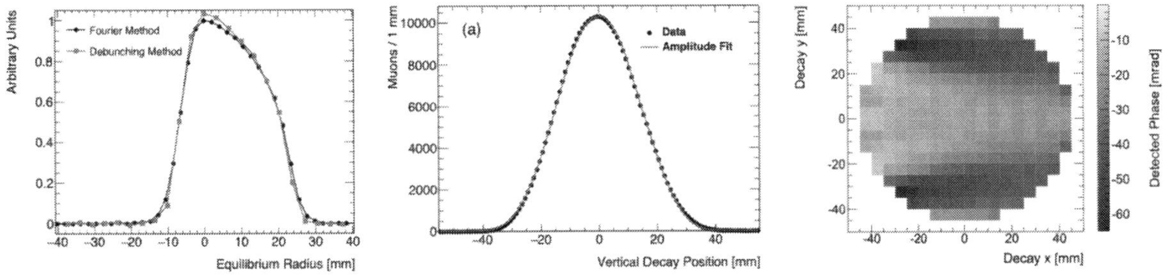

Figure 4. Left panel: The measured equilibrium radius determined with two methods. Middle panel: The vertical distribution of decay positron measured with the straw detectors. Right panel: The simulated phase map for decay positrons over the storage aperture.

of tracker data and simulated maps of the positron acceptance, asymmetry, and phase [9]. An example of the simulated phase map is shown in the right panel of Fig. 4. The correction applied in Run-1 was $C_{pa} = -158 \pm 75$, the second largest systematic uncertainty.

3.4 Measuring $\langle \omega_p(x, y, \phi) \times M(x, y, \phi) \rangle$

The measurement of the magnetic field in terms of the Larmor frequency of shielded protons weighted by the muon beam distribution, $\langle \omega_p(x, y, \phi) \times M(x, y, \phi) \rangle$, is one of the key ingredients in the denominator of Eq. (4). The term $\omega_p(x, y, \phi)$ is derived from the two main NMR systems in the experiment, namely the 378 fixed probes and the trolley system, the latter being shown in the left panel of Fig 5. Every 3 to 5 days, the muon beam injection is stopped and the trolley takes a full field map with about 9000 measurements for each of its 17 probes around the full 360° of the storage ring. The continuous measurements of the 378 fixed probes are synchronized to the detailed field maps taken by the trolley. Between trolley maps, the fixed probes then monitor the slow field drift. As they are arranged in groups of 4 and 6 probes in 72 azimuthal locations, they also provide some information about the higher order multipole drift. Both the detailed trolley measurements and the fixed probe information are analyzed using various methods including a Hilbert transform method [66] and are then combined in a complex analysis chain, to generate the frequency maps $\omega_p(x, y, \phi)$ [11].

The straw trackers provide the information needed to extract the muon distribution maps $M(x, y, \phi)$ over the course of the data taking. To extract the relevant information, the full analysis takes into account beam dynamics effects (such as beta functions) and detector acceptances of the trackers and the calorimeters.

3.5 Magnetic field transients

The pulsed injection systems used in the experiment, the fast magnetic kicker and the electric quadrupoles create magnetic transient fields that are too fast to be measurable with the main pulsed NMR probes employed. To investigate the magnetic field originating from the kicker, two Faraday magnetometers were built, one of them based on the magnetometer from E821 [67]. Details about magnetometer measurements are presented in a separate contribution to this conference.

In normal operating mode, the fixed probes are constantly taking measurements at ~1 Hz rate asynchronously with the beam injection. A special running mode was setup to trigger them relative to the beam injection with varying delay time. These special studies revealed a sizable change in the magnetic field which is now believed to be understood in terms of the mechanical vibrations of the quadrupole plates inducing magnetic fields. Since the NMR probes have a low bandwidth and are shielded by the vacuum chambers from high frequencies, a dedicated probe setup was developed to measure the effect in the storage region inside the quadrupoles. It was used to provide a detailed measurement of the change in the magnetic field before, during, and after muon injection. Fig 6 shows the sizable magnetic field change in the vicinity of one of the muon storage cycles. During the 700 μs of muon storage (gray band), the quadrupole induced field transient yields a change in the magnetic field of 100 – 200 ppb. A more detailed analysis [11] using measurements at various azimuthal and radial positions, averaging over all 16 different muon injection cycles, and folding the actual field change with the change in stored muons during the 700 μs yields a correction of $B_q = -17 \pm 92$, the single largest systematic uncertainty for the Run-1 result.

4 Summary and Outlook

The Run-1 result was based on about 6% of the total statistics that the experiment has collected and led to a confirmation of the BNL measurement, increasing the tension with the SM to $4.2\,\sigma$. Over the course of 2019 to 2023, a lot more data was collected. Fig. 7 shows the rate of data collection for all data taking periods (Run-1 through Run-6) in terms of the number of total BNL statistics. The technical design report's goal [7] of collecting 21 more statistics than BNL was reached in February 2023. Data analysis for the Run-2/3 datasets is close to final and on track to reduce the statistical uncertainty of the first result by a factor of 2 and bring the total systematic uncertainty to ~100 ppb. The unblinding of this ongoing analysis is expected in the late spring of 2023. The remaining datasets from Run-4, Run-5, and Run-6 will bring an-

Figure 5. Left panel: The in-vacuum field camera, the so-called trolley. Right panel: The azimuthally averaged magnetic field contours $\omega_p(x, y)$ overlaid with the azimuthally averaged muon distribution $M(x, y)$.

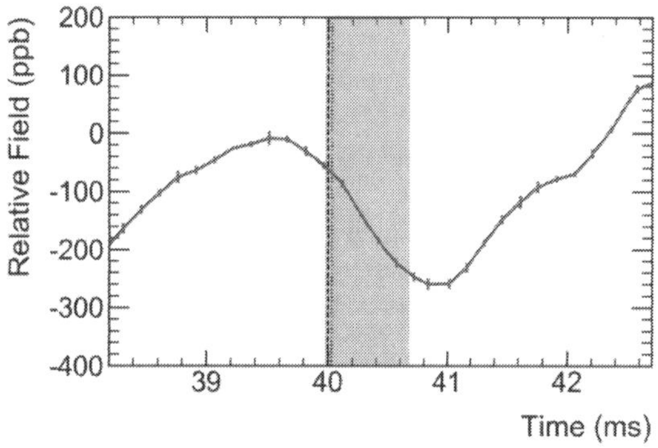

Figure 6. Measured change in field due to the quadrupole plate vibrations for one of the 16 muon injection cycles.

other factor of 2 for the total statistical uncertainty and some further reduction of the total systematic, leading to the anticipated overall factor of 4 improvement compared to BNL. After the first data taking period, a few key upgrades were performed that will help to reduce some of the systematic uncertainties compared to the Run-1 result. These upgrades include i) new qaudrupole resistors replacing the broken ones which will significantly reduce C_{pa}, ii) modifications of the kicker system to increase the voltage and center the beam radially, iii) thermal magnet insulation and better hall cooling to reduce the field drift, iv) improved quadrupole transient measurements significantly reducing the largest systematic uncertainty in Run-1, and v) a new RF system [68] which suppresses the coherent betatron oscillation amplitude and its associated systematic uncertainty. A final result for the Run-4/5/6 datasets is anticipated about 2-3 years from now.

5 Acknowledgments

This research was supported by the U.S. Department of Energy, Office of Science, High Energy Physics under contracts DE-AC02-06CH11357 (Argonne National Labo- ratory), and by Fermi National Accelerator Laboratory (Fermilab), a US DOE, HEP User Facility. Fermilab is managed by Fermi Research Alliance, LLC (FRA), acting under Contract No. DE-AC02-07CH11359.

References

[1] P.A.M. Dirac, Proc. R. Soc. A (London) **117**, 610 (1928)

[2] J.S. Schwinger, Phys. Rev. **73**, 416 (1948)

[3] X. Fan, T.G. Myers, B.A.D. Sukra, G. Gabrielse, Phys. Rev. Lett. **130**, 071801 (2023)

[4] R.H. Parker, C. Yu, W. Zhong, B. Estey, H. Müller, Science **360**, 191 (2018)

[5] L. Morel, Z. Yao, P. Cladé, S. Guellati-Khélifa, Nature **588**, 61 (2020)

[6] G.W. Bennett, B. Bousquet, H.N. Brown, G. Bunce, R.M. Carey, P. Cushman, G.T. Danby, P.T. Debevec, M. Deile, H. Deng et al. (Muon g-2), Phys. Rev. D **73**, 072003 (2006), `hep-ex/0602035`

[7] J. Grange, V. Guarino, P. Winter, K. Wood, H. Zhao, R.M. Carey, D. Gastler, E. Hazen, N. Kinnaird, J.P. Miller et al. (Muon g-2), arXiv:1501.06858 [physics.ins-det] (2015), `1501.06858`

[8] B. Abi, T. Albahri, S. Al-Kilani, D. Allspach, L.P. Alonzi, A. Anastasi, A. Anisenkov, F. Azfar, K. Badgley, S. Baeßler et al. (Muon $g-2$), Phys. Rev. Lett **126**, 141801 (2021)

[9] T. Albahri, A. Anastasi, K. Badgley, S. Baeßler, I. Bailey, V.A. Baranov, E. Barlas-Yucel, T. Barrett, F. Bedeschi, M. Berz et al. (Muon $g-2$), Phys. Rev. Accel. Beams **24**, 044002 (2021)

[10] T. Albahri, A. Anastasi, A. Anisenkov, K. Badgley, S. Baeßler, I. Bailey, V.A. Baranov, E. Barlas-Yucel, T. Barrett, A. Basti et al. (Muon $g-2$), Phys. Rev. D **103**, 072002 (2021)

[11] T. Albahri, A. Anastasi, K. Badgley, S. Baeßler, I. Bailey, V.A. Baranov, E. Barlas-Yucel, T. Barrett, F. Bedeschi, M. Berz et al. (Muon $g-2$), Phys. Rev. A **103**, 042208 (2021)

[12] T. Aoyama, N. Asmussen, M. Benayoun, J. Bijnens, T. Blum, M. Bruno, I. Caprini, C.M. Carloni Calame, M. Cè, G. Colangelo et al., Phys. Rep. **887**, 1 (2020)

[13] T. Aoyama, M. Hayakawa, T. Kinoshita, M. Nio, Phys. Rev. Lett. **109**, 111808 (2012), `1205.5370`

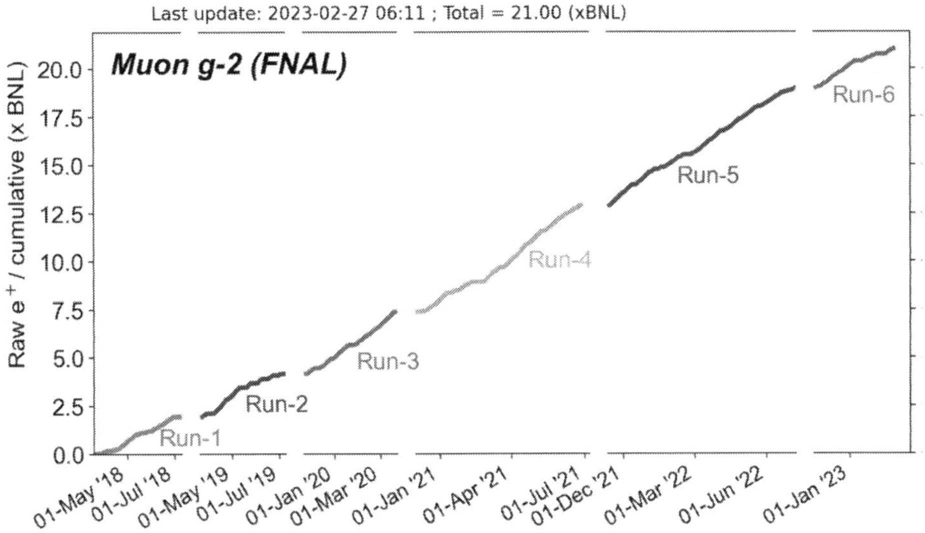

Figure 7. Raw collected statistics for all data taking periods. The technical design goal of 21× BNL statistics was reached in 02/2023.

[14] T. Aoyama, T. Kinoshita, M. Nio, Atoms **7**, 28 (2019)

[15] A. Czarnecki, W.J. Marciano, A. Vainshtein, Phys. Rev. **D67**, 073006 (2003), [Erratum: Phys. Rev. **D73**, 119901 (2006)], hep-ph/0212229

[16] C. Gnendiger, D. Stöckinger, H. Stöckinger-Kim, Phys. Rev. D **88**, 053005 (2013), 1306.5546

[17] M. Davier, A. Hoecker, B. Malaescu, Z. Zhang, Eur. Phys. J. **C77**, 827 (2017), 1706.09436

[18] A. Keshavarzi, D. Nomura, T. Teubner, Phys. Rev. **D97**, 114025 (2018), 1802.02995

[19] G. Colangelo, M. Hoferichter, P. Stoffer, J. High Energy Phys. **02**, 006 (2019), 1810.00007

[20] M. Hoferichter, B.L. Hoid, B. Kubis, J. High Energy Phys. **08**, 137 (2019), 1907.01556

[21] M. Davier, A. Hoecker, B. Malaescu, Z. Zhang, Eur. Phys. J. **C80**, 241 (2020), 1908.00921

[22] A. Keshavarzi, D. Nomura, T. Teubner, Phys. Rev. D **101**, 014029 (2020), 1911.00367

[23] A. Kurz, T. Liu, P. Marquard, M. Steinhauser, Phys. Lett. **B734**, 144 (2014), 1403.6400

[24] K. Melnikov, A. Vainshtein, Phys. Rev. D **70**, 113006 (2004), hep-ph/0312226

[25] P. Masjuan, P. Sánchez-Puertas, Phys. Rev. D **95**, 054026 (2017), 1701.05829

[26] G. Colangelo, M. Hoferichter, M. Procura, P. Stoffer, J. High Energy Phys. **04**, 161 (2017), 1702.07347

[27] M. Hoferichter, B.L. Hoid, B. Kubis, S. Leupold, S.P. Schneider, J. High Energy Phys. **10**, 141 (2018), 1808.04823

[28] A. Gérardin, H.B. Meyer, A. Nyffeler, Phys. Rev. D **100**, 034520 (2019), 1903.09471

[29] J. Bijnens, N. Hermansson-Truedsson, A. Rodríguez-Sánchez, Phys. Lett. B **798**, 134994 (2019), 1908.03331

[30] G. Colangelo, F. Hagelstein, M. Hoferichter, L. Laub, P. Stoffer, J. High Energy Phys. **03**, 101 (2020), 1910.13432

[31] T. Blum, N. Christ, M. Hayakawa, T. Izubuchi, L. Jin, C. Jung, C. Lehner, Phys. Rev. Lett. **124**, 132002 (2020), 1911.08123

[32] G. Colangelo, M. Hoferichter, A. Nyffeler, M. Passera, P. Stoffer, Phys. Lett. **B735**, 90 (2014), 1403.7512

[33] F.V. Ignatov, R.R. Akhmetshin, A.N. Amirkhanov, A.V. Anisenkov, V.M. Aulchenko, N.S. Bashtovoy, D.E. Berkaev, A.E. Bondar, A.V. Bragin, S.I. Eidelman et al. (CMD-3 Collaboration), *Measurement of the $e^+e^- \to \pi^+\pi^-$ cross section from threshold to 1.2 GeV with the CMD-3 detector* (2023), 2302.08834

[34] S. Borsanyi, Z. Fodor, J.N. Guenther, C. Hoelbling, S.D. Katz, L. Lellouch, T. Lippert, K. Miura, L. Parato, K.K. Szabo et al., Nature **593**, 51 (2021), 2002.12347

[35] C. Lehner, A.S. Meyer, Physical Review D **101**, 074515 (2020), 2003.04177

[36] G. Wang, T. Draper, K.F. Liu, Y.B. Yang, *Muon g-2 with overlap valence fermions* (2022), https://arxiv.org/abs/2204.01280

[37] C. Aubin, T. Blum, M. Golterman, S. Peris, Physical Review D **106**, 054503 (2022), 2204.12256

[38] M. Cè, A. Gérardin, G.v. Hippel, R.J. Hudspith, S. Kuberski, H.B. Meyer, K. Miura, D. Mohler, K. Ottnad, S. Paul et al., Physical Review D **106**, 114502 (2022), 2206.06582

[39] C. Alexandrou, S. Bacchio, P. Dimopoulos, J. Finkenrath, R. Frezzotti, G. Gagliardi, M. Garofalo, K. Hadjiyiannakou, B. Kostrzewa, K. Jansen et al., *Lattice calculation of the short and intermediate time-distance hadronic vacuum polarization contributions to the muon magnetic moment using twisted-mass fermions* (2022), https://arxiv.org/abs/2206.15084

[40] T. Blum, P.A. Boyle, M. Bruno, D. Giusti, V. Gülpers, R.C. Hill, T. Izubuchi, Y.C. Jang, L. Jin, C. Jung et al., *An update of Euclidean windows of the hadronic vacuum polarization* (2023), https://arxiv.org/abs/2301.08696

[41] A. Bazavov, C. Davies, C. DeTar, A.X. El-Khadra, E. Gámiz, S. Gottlieb, W.I. Jay, H. Jeong, A.S. Kronfeld, S. Lahert et al., *Light-quark connected intermediate-window contributions to the muon $g-2$ hadronic vacuum polarization from lattice QCD* (2023), https://arxiv.org/abs/2301.08274

[42] G. Charpak, F.J.M. Farley, R.L. Garwin, Phys. Lett. **1**, 16 (1962)

[43] J. Bailey, W. Bartl, G. von Bochmann, R.C.A. Brown, F.J.M. Farley et al., Nuovo Cim. A **9**, 369 (1972)

[44] J. Bailey et al. (CERN-Mainz-Daresbury), Nucl. Phys. B **150**, 1 (1979)

[45] A. Yamamoto et al., Nucl. Instrum. Meth. **A491**, 23 (2002)

[46] A. Schreckenberger, D. Allspach, D. Barak, J. Bohn, C. Bradford, D. Cauz, S. Chang, A. Chapelain, S. Chappa, S. Charity et al., Nuclear Instruments and Methods in Physics Research Section A: Accelerators, Spectrometers, Detectors and Associated Equipment **1011**, 165597 (2021)

[47] Y.K. Semertzidis, G. Bennett, E. Efstathiadis, F. Krienen, R. Larsen, Y.Y. Lee, W.M. Morse, Y. Orlov, C.S. Ozben, B.L. Roberts et al., Nucl. Instrum. Methods Phys. Res., Sect. A **503**, 458 (2003)

[48] M. Abe, S. Bae, G. Beer, G. Bunce, H. Choi, S. Choi, M. Chung, W. da Silva, S. Eidelman, M. Finger et al., Progress of Theoretical and Experimental Physics **2019** (2019)

[49] A.T. Fienberg, L.P. Alonzi, A. Anastasi, R. Bjorkquist, D. Cauz, R. Fatemi, C. Ferrari, A. Fioretti, A. Frankenthal, C. Gabbanini et al., Nucl. Instrum. Methods Phys. Res., Sect. A **783**, 12 (2015), 1412.5525

[50] J. Kaspar, A.T. Fienberg, D.W. Hertzog, M.A. Huehn, P. Kammel, K.S. Khaw, D.A. Peterson, M.W. Smith, T.D.V. Wechel, A. Chapelain et al., J. Instrum. **12**, P01009 (2017), 1611.03180

[51] K.S. Khaw, M. Bartolini, H. Binney, R. Bjorkquist, A. Chapelain, A. Driutti, C. Ferrari, A.T. Fienberg, A. Fioretti, C. Gabbanini et al. (Muon g-2), Nucl. Instrum. Methods Phys. Res., Sect. A **945**, 162558 (2019), 1905.04407

[52] A. Anastasi, A. Basti, F. Bedeschi, A. Boiano, E. Bottalico, G. Cantatore, D. Cauz, A.T. Chapelain, G. Corradi, S. Dabagov et al. (Muon g-2), J. Instrum. **14**, P11025 (2019), 1906.08432

[53] F. Bloch, W.W. Hansen, M. Packard, Phys. Rev. **69**, 127 (1946)

[54] E.M. Purcell, H.C. Torrey, R.V. Pound, Phys. Rev. **69**, 37 (1946)

[55] S. Corrodi, P.D. Lurgio, D. Flay, J. Grange, R. Hong, D. Kawall, M. Oberling, S. Ramachandran, P. Winter, J. Instrum. **15**, P11008 (2020)

[56] D. Flay, D. Kawall, T. Chupp, S. Corrodi, M. Farooq, M. Fertl, J. George, J. Grange, R. Hong, R. Osofsky et al., Journal of Instrumentation **16**, P12041 (2021), 2109.08992

[57] M. Farooq, T. Chupp, J. Grange, A. Tewsley-Booth, D. Flay, D. Kawall, N. Sachdeva, P. Winter, Phys. Rev. Lett. **124**, 223001 (2020)

[58] B. King, T. Albahri, S. Al-Kilani, D. Allspach, D. Beckner, A. Behnke, T. Bowcock, D. Boyden, R. Carey, J. Carroll et al., Journal of Instrumentation **17**, P02035 (2022)

[59] W.D. Phillips, W.E. Cooke, D. Kleppner, Metrologia **13**, 179 (1977)

[60] P.J. Mohr, D.B. Newell, B.N. Taylor, Rev. Mod. Phys. **88**, 035009 (2016), 1507.07956

[61] W. Liu, M.G. Boshier, S. Dhawan, O. van Dyck, P. Egan, X. Fei, M.G. Perdekamp, V.W. Hughes, M. Janousch, K. Jungmann et al., Phys. Rev. Lett. **82**, 711 (1999)

[62] Y. Orlov, C.S. Ozben, Y.K. Semertzidis, Nuclear Instruments and Methods in Physics Research Section A: Accelerators, Spectrometers, Detectors and Associated Equipment **482**, 767 (2002)

[63] F. Combley, E. Picasso, Physics Reports **14**, 1 (1974)

[64] F. Farley, Physics Letters B **42**, 66 (1972)

[65] F.J.M. Farley, E. Picasso, Adv. Ser. Direct. High Energy Phys. **7**, 479 (1990)

[66] R. Hong, S. Corrodi, S. Charity, S. Baessler, J. Bono, T. Chupp, M. Fertl, D. Flay, A. Garcia, J. George et al. (Muon $g-2$), Journal of Magnetic Resonance **329**, 107020 (2021), 2101.08412

[67] E. Efstathiadis, Y.Y. Lee, J.L. Mi, C. Pai, J.M. Paley, B.L. Roberts, R.T. Sanders, Y.K. Semertzidis, D.S. Warburton, Nucl. Instrum. Methods Phys. Res., Sect. A **496**, 8 (2003)

[68] O. Kim, M. Bhattacharya, S. Chang, J. Choi, J.D. Crnkovic, S. Ganguly, S. Haciömeroğlu, M. Kargiantoulakis, Y.I. Kim, S. Lee et al., New Journal of Physics **22**, 063002 (2020)

Model independent analysis for B anomalies

Farvah Mahmoudi[1,2,*], Tobias Hurth[3], Diego Martínez Santos[4], and Siavash Neshatpour[5]

[1] *Université de Lyon, Université Claude Bernard Lyon 1, CNRS/IN2P3, Institut de Physique des 2 Infinis de Lyon, UMR 5822, F-69622, Villeurbanne, France*
[2] *CERN, Theoretical Physics Department, CH-1211 Geneva 23, Switzerland*
[3] *PRISMA+ Cluster of Excellence and Institute for Physics (THEP), Johannes Gutenberg University, D-55099 Mainz, Germany*
[4] *Instituto Galego de Física de Altas Enerxías, Universidade de Santiago de Compostela, Spain*
[5] *INFN-Sezione di Napoli, Complesso Universitario di Monte S. Angelo, Via Cintia Edificio 6, 80126 Napoli, Italy*

Abstract. We discuss the implications of $b \to s\ell^+\ell^-$ measurements and their deviations with respect to the Standard Model predictions in a model-independent framework. We highlight in particular the impact of the recent updated measurements including the updated $B_s \to \phi\mu^+\mu^-$ branching ratios and angular observables, the recent CMS measurement of the branching ratio of $B_s \to \mu^+\mu^-$, and the LHCb measured lepton flavour universality violating ratios $R_{K_S^0}$ and $R_{K^{*+}}$. In addition, we check the compatibility of the new physics effect for the theoretically clean observables with the rest of the neutral B decays observables.

1 Introduction

In the last few years, since the measured deviation in the angular observable P_5' of the $B \to K^*\mu^+\mu^-$ decay [1], there have been several measurements in neutral B-decays indicating tension with the Standard Model (SM). Updated measurements by LHCb for the $P_5'(B \to K^*\mu^+\mu^-)$ have persistently shown tension with the SM which can be explained with short distance new physics (NP) contributions [2, 3]. This is also the case of the overall $B \to K^*\mu^+\mu^-$ angular observables and is supported in addition (see e.g. [4]) by the angular analysis of its isospin partner in the recent measurement of $B^+ \to K^{*+}\mu^+\mu^-$ [5]. The $B_s \to \phi\mu^+\mu^-$ branching fraction [6–8] also indicates tensions with the SM and is measured to be below the SM prediction. This trend is seen in several other $b \to s\ell^+\ell^-$ branching fractions such as $B \to K\mu^+\mu^-$ [9] and $\Lambda_b \to \Lambda\mu^+\mu^-$ [10]. Since the branching fractions are dependent on the relevant local form factors, they suffer from large theoretical uncertainties. In contrast, the angular observables have a reduced sensitivity to the form factor uncertainties, but they are still dependent on the non-local hadronic contributions whose size are not fully known in QCD factorisation. Consequently the significance of the anomalies are dependent on the estimated size of the non-local effects. Recent theoretical progress for a better control of these effects can be found in Refs. [11–13].

A set of observables to test lepton flavour universality violation (LFUV) in $b \to s\ell^+\ell^-$ transitions is defined as $R_H = (B \to H\mu^+\mu^-)/(B \to He^+e^-)$ with $H = K^+, K^*, \phi, ...$ [14]. Unlike the observables mentioned in the previous paragraph, these ratios are very precisely known in the SM. There have been signs of deviation from the SM in the LFUV ratios for the case of R_K [15–17] and R_{K^*} [18]. The recent measurements of $R_{K_S^0}$ and $R_{K^{*+}}$ [19] although within 2σ of the SM prediction, show the same trend as their isospin partners with the central values below the SM predictions. Incidentally, there have also been a slight sign of LFUV in flavour changing neutral current processes in the Kaon sector [20] (currently the ex-

*e-mail: nazila@cern.ch

https://doi.org/10.1051/epjconf/202328901002

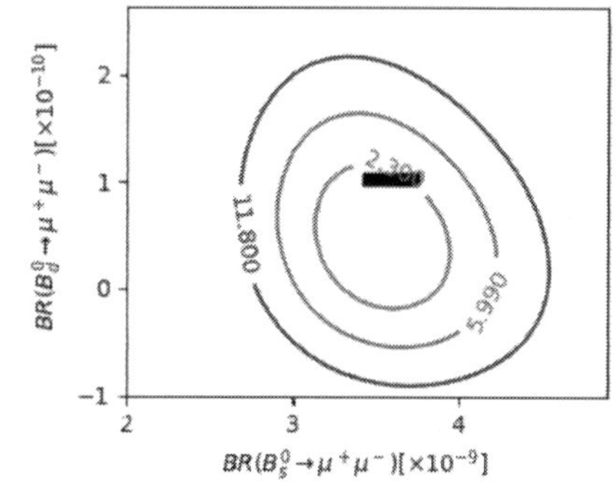

Figure 1. Two dimensional likelihood plot of BR($B_{s,d} \to \mu^+\mu^-$).

perimental uncertainty is quite large for these processes).

The significance of each of the B-anomalies, individually is around $\sim 2-3\sigma$, however collectively they can be explained by common NP scenarios and have a much larger significance in a global analysis [21–26].

Another precisely predicted observable with an uncertainty of less than 5% in its SM prediction is BR($B_s \to \mu^+\mu^-$) which has been measured by several experiments. Previous measurements of ATLAS [27], CMS [28] and LHCb [29, 30] were in about 2σ tension with the SM prediction [21]. However, the situation has changed with the recent data from CMS [31]. For our fits, we combined the ATLAS [27] and LHCb [29, 30] results together with the recent CMS [31] measurement, considering a joint 2D likelihood as shown in Fig 1. We obtained the experimental combined value of the $B_s \to \mu^+\mu^-$ branching ratio to be

$$\mathrm{BR(B_s} \to \mu^+\mu^-)^{\text{comb.}}_{\text{exp}} = \left(3.52^{+0.32}_{-0.30}\right) \times 10^{-9}, \quad (1)$$

which is within 1σ of the SM prediction.

2 Coherence of clean observables with the rest of the rare *B*-decay observables

In order to examine the consistency of the implication of the clean observables for new physics as compared to the rest of the observables [32, 33], we perform two sets of fits; one to the clean observables where we consider R_K, R_{K^*} as well as their isospin partners $R_{K^0_S}$ and $R_{K^{*+}}$ [19] and also BR($B_{s,d} \to \mu^+\mu^-$), and another one considering the rest of the $bs\ell\ell$ observables. The observable calculations and the χ^2 fitting is done using the SuperIso public program [34–38].

2.1 Clean observables

In table 1 we give the one-dimensional NP fits to clean observables and compare them with our 2021 fit results [21]. Compared to Ref. [21], we now include the two LFUV ratios $R_{K^0_S}$ and $R_{K^{*+}}$ [19] as well as the R_K measurement by Belle [39] in the [1,6] GeV^2 bin and the updated combination for BR($B_s \to \mu^+\mu^-$) as given in Eq. (1).

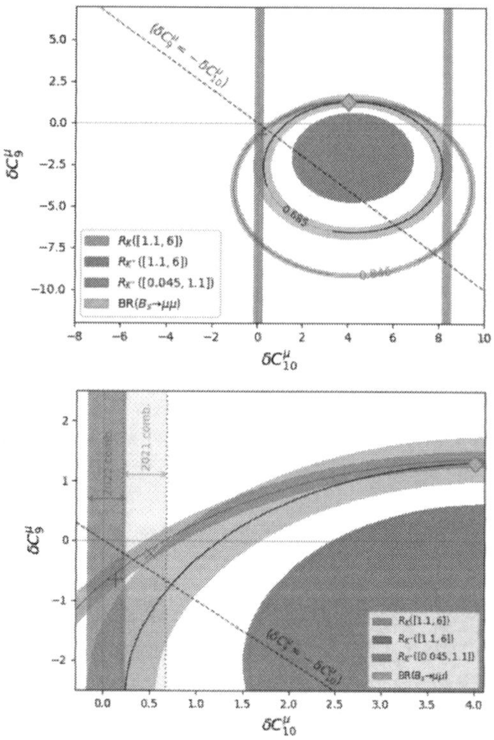

Figure 2. The prediction of $R_{K^{(*)}}$ and BR($B_s \to \mu^+\mu^-$) within $1\sigma\left(= \sqrt{\sigma^2_{\text{th}} + \sigma^2_{\text{exp}}}\right)$ of their measured values. On the lower plot we have the zoomed-in version of the upper plot. The dark gray band indicates the 1σ region corresponding to the updated combination of BR($B_s \to \mu\mu$) and the lighter gray region (on the lower plot) with the dotted borders corresponds to the 2021 combination. The yellow diamond indicates the best fit value to $R_{K^{(*)}}$, while the green plus sign (gray cross) corresponds to the best fit point when the 2022 (2021) combination for BR($B_s \to \mu\mu$) is included in the fit.

https://doi.org/10.1051/epjconf/202328901002

Only LFUV ratios and $B_{s,d} \to \ell^+\ell^-$ **2021 fit results** ($\chi^2_{\rm SM} = 28.19$)			
	b.f. value	$\chi^2_{\rm min}$	Pull$_{\rm SM}$
δC_9	-1.00 ± 6.00	28.1	0.2σ
δC_9^e	0.80 ± 0.21	11.2	4.1σ
δC_9^μ	-0.77 ± 0.21	11.9	4.0σ
δC_{10}	0.43 ± 0.24	24.6	1.9σ
δC_{10}^e	-0.78 ± 0.20	9.5	4.3σ
δC_{10}^μ	0.64 ± 0.15	7.3	4.6σ
$\delta C_{\rm LL}^e$	0.41 ± 0.11	10.3	4.2σ
$\delta C_{\rm LL}^\mu$	-0.38 ± 0.09	7.1	4.6σ

Only LFUV ratios and $B_{s,d} \to \ell^+\ell^-$ **2022 fit results** ($\chi^2_{\rm SM} = 30.63$)			
	b.f. value	$\chi^2_{\rm min}$	Pull$_{\rm SM}$
δC_9	-2.00 ± 5.00	30.5	0.4σ
δC_9^e	0.83 ± 0.21	10.8	4.4σ
δC_9^μ	-0.80 ± 0.21	11.8	4.3σ
δC_{10}	0.03 ± 0.20	30.6	0.1σ
δC_{10}^e	-0.81 ± 0.19	8.7	4.7σ
δC_{10}^μ	0.50 ± 0.14	16.2	3.8σ
$\delta C_{\rm LL}^e$	0.43 ± 0.11	9.7	4.6σ
$\delta C_{\rm LL}^\mu$	-0.33 ± 0.08	12.4	4.3σ

Table 1. Comparison of the fits to clean observables with the 2021 fit results [21] on the left and the updated 2022 fits on the right.

All observables except LFUV ratios and $B_{s,d} \to \ell^+\ell^-$ **2021 fit results** ($\chi^2_{\rm SM} = 200.1$)			
	b.f. value	$\chi^2_{\rm min}$	Pull$_{\rm SM}$
δC_9	-1.01 ± 0.13	158.2	6.5σ
δC_9^e	0.70 ± 0.60	198.8	1.1σ
δC_9^μ	-1.03 ± 0.13	156.0	6.6σ
δC_{10}	0.34 ± 0.23	197.7	1.5σ
δC_{10}^e	-0.50 ± 0.50	199.0	1.0σ
δC_{10}^μ	0.41 ± 0.23	196.5	1.9σ
$\delta C_{\rm LL}^e$	0.33 ± 0.29	198.9	1.1σ
$\delta C_{\rm LL}^\mu$	-0.75 ± 0.13	167.9	5.7σ

All observables except LFUV ratios and $B_{s,d} \to \ell^+\ell^-$ **2022 fit results** ($\chi^2_{\rm SM} = 221.8$)			
	b.f. value	$\chi^2_{\rm min}$	Pull$_{\rm SM}$
δC_9	-0.95 ± 0.13	185.1	6.1σ
δC_9^e	0.70 ± 0.60	220.5	1.1σ
δC_9^μ	-0.96 ± 0.13	182.8	6.2σ
δC_{10}	0.29 ± 0.21	219.8	1.4σ
δC_{10}^e	-0.60 ± 0.50	220.6	1.1σ
δC_{10}^μ	0.35 ± 0.20	218.7	1.8σ
$\delta C_{\rm LL}^e$	0.34 ± 0.29	220.6	1.1σ
$\delta C_{\rm LL}^\mu$	-0.64 ± 0.13	195.0	5.2σ

Table 2. Comparison of the fits to all observables except the clean ones with the 2021 fit results on the left and the updated 2022 fits on the right.

While the significance of NP in $C_9^{e,\mu}$ or C_{10}^e has slightly increased, the C_{10}^μ solution is now less favoured compared to the 2021 results [21]. This is expected as the new combination of BR($B_s \to \mu^+\mu^-$) is now in much better agreement with the SM prediction and constrains more C_{10}^μ. The inclusion of BR($B_s \to \mu^+\mu^-$) in this set of observables is crucial in breaking the degeneracy between NP in δC_9^μ and δC_{10}^μ for explaining the measured values of the LFUV ratios as can be clearly seen in figure 2 where without BR($B_s \to \mu^+\mu^-$) the best fit point of $R_{K^{(*)}}$ is given by the yellow diamond while including it moves the best fit value to the green plus sign. The impact of the updated value of the BR($B_s \to \mu^+\mu^-$) can bee seen in the lower plot by comparing the green plus sign with the gray cross corresponding to the best fit point when the 2021 combination for BR($B_s \to \mu^+\mu^-$) was considered.

2.2 All except the clean observables

We consider now the 1-dimensional NP fits to the rest of the observables, excluding the LFUV ratios and $B_{s,d} \to \ell^+\ell^-$. We assume 10% power correction for the non-factorisable contributions beyond QDC factorisation [40–43]. Compared to Ref. [21] we use the updated LHCb results for the $B_s \to \phi\mu^+\mu^-$ observables [7, 8] with 8.4 fb^{-1} of data. The CMS measurement for $F_H(B^+ \to K^+\mu^+\mu^-)$ [44] and the LHCb measurement of the angular observables of $B \to K^*e^+e^-$ [45] have also been considered. As can be seen in table 2, the hierarchy of the preferred NP contributions is similar to the 2021 results, where the most preferred scenarios are still NP in lepton flavour violating δC_9^μ and NP in lepton flavour universal δC_9 with the third most preferred description given by NP in the chiral basis $\delta C_{\rm LL}^\mu$. The above mentioned scenarios however are show-

 https://doi.org/10.1051/epjconf/202328901002

ing a $\sim 0.4\sigma$ reduced significance compared to our 2021 results which is mainly due to the updated $B_s \to \phi\mu^+\mu^-$ experimental data.

For the fit to all observables except the clean ones there is no significant indication for NP within the electron sector since not only the measurements in the electron sector are in good agreement with their SM predictions, there are also far less data compared to the decays with muons. Comparing the result of table 2 with the result of the previous subsection (table 1) we see that there is not a full agreement for the preferred scenarios, however, there are common scenarios such as NP contributions to δC_9^μ [46, 47] which have a large significance for both datasets with best fit points that agree within 1σ.

The compatibility of the two-dimensional NP fits to "clean observables" and the NP fits to "all observables except the clean ones" can be seen in figure 3 where also the significant impact of including or removing BR($B_s \to \mu^+\mu^-$) from each dataset is clearly visible, especially for the clean observables.

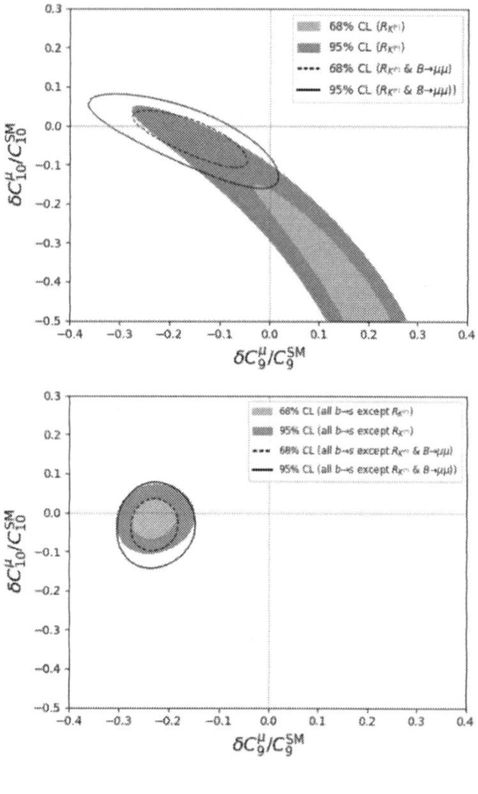

Figure 3. Two-dimensional fits to the clean observables (top) and to the rest of the observables (bottom).

3 Global fits to all $b \to s\ell^+\ell^-$ observables

For a global analysis of the NP implications of rare B-decays, we need to take into account all relevant $b \to s$ decays combining the datasets of section 2.1 and section 2.2. We assume 10% error for the power corrections when applicable.

3.1 One- and two-dimensional fits

The 1-dim NP fits to the rare B-decays are given in table 3. As anticipated from the comparison of the fits to clean observables and the rest of the observables, the most favoured scenario to explain the overall data is lepton flavour violating NP in δC_9^μ. The other prominent scenarios are NP in $\delta C_{\rm LL}^\mu$ followed by lepton flavour universal NP in δC_9. While the hierarchy of the favoured scenarios has not changed, it should be noted that NP in δC_{10}^μ is now less favoured which is mostly due to the updated combination for BR($B_s \to \mu^+\mu^-$). This can also be seen in the decrease of the significance of δC_{LL}^μ compared to the 2021 fit results. The decrease of preference of NP in δC_{10}^μ can also be seen in the 2-dim fits of figure 4.

In the 1-and 2-dim fits of table 3 and figure 4 we have not shown the NP fits to the radiative coefficient δC_7, the scalar and pseudoscalar coefficients ($\delta C_{Q_{1,2}}$) or the coefficients where the hadronic currents are right-handed ($\delta C_i'$) since they are all strongly constrained by data. The situation can in principle change when several coefficients can simultaneously contribute, this is clearly the case when doing a simultaneous fit to δC_{10}^μ and $\delta C_{Q_{1,2}}^\mu$ [48] which would otherwise be severely constrained by BR($B_s \to \mu^+\mu^-$) if only one single coefficient would contribute.

3.2 Multidimensional fit

A multidimensional fit gives in principle a more realistic picture than assuming new physics contribution to only a single coefficient, as it is very unlikely for a UV-complete scenario to merely affect one coefficient while the rest of the coefficients are kept to their SM values. Therefore, here we consider a 20-dim fit varying all relevant Wilson coefficients (table 4). Besides being more realistic, this multidimensional fit has the advantage of avoiding the look elsewhere effect (LEE) since LEE not only takes place when one makes a selected choice of observables but is also relevant in the case when a posteriori a subset of specific NP directions are

https://doi.org/10.1051/epjconf/202328901002

All observables **2021 fit results** ($\chi^2_{\rm SM} = 225.8$)			
	b.f. value	$\chi^2_{\rm min}$	Pull$_{\rm SM}$
δC_9	-0.99 ± 0.13	186.2	6.3σ
δC_9^e	0.79 ± 0.20	207.7	4.3σ
δC_9^μ	-0.95 ± 0.12	168.6	7.6σ
δC_{10}	0.32 ± 0.18	222.3	1.9σ
δC_{10}^e	-0.74 ± 0.18	206.3	4.4σ
δC_{10}^μ	0.55 ± 0.13	205.2	4.5σ
$\delta C_{\rm LL}^e$	0.40 ± 0.10	206.9	4.3σ
$\delta C_{\rm LL}^\mu$	-0.49 ± 0.08	180.5	6.7σ

All observables **2022 fit results** ($\chi^2_{\rm SM} = 253.5$)			
	b.f. value	$\chi^2_{\rm min}$	Pull$_{\rm SM}$
δC_9	-0.95 ± 0.13	215.8	6.1σ
δC_9^e	0.82 ± 0.19	232.4	4.6σ
δC_9^μ	-0.92 ± 0.11	195.2	7.6σ
δC_{10}	0.08 ± 0.16	253.2	0.5σ
δC_{10}^e	-0.77 ± 0.18	230.6	4.8σ
δC_{10}^μ	0.43 ± 0.12	238.9	3.8σ
$\delta C_{\rm LL}^e$	0.42 ± 0.10	231.4	4.7σ
$\delta C_{\rm LL}^\mu$	-0.43 ± 0.07	213.6	6.3σ

Table 3. Comparison of the fits to all observables with the 2021 fit results on the left and the updated 2022 fits on the right.

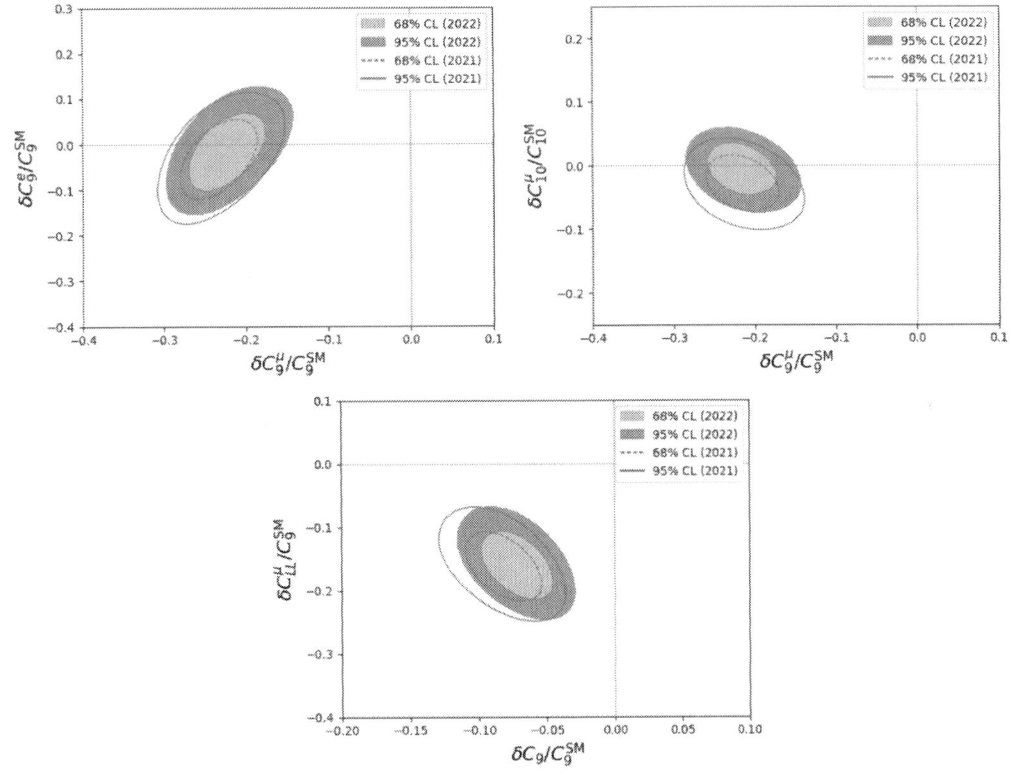

Figure 4. Two-dimensional fit to all rare B-decay observables.

assumed which is circumvented when all possible Wilson coefficients are varied. With a large set of free parameters and the limited decay modes there can be flat directions or non-sensitive NP coefficients that can be removed by considering the correlations and likelihood profiles in order to get an "effective" number of degrees of freedom (dof$_{\rm eff}$). In the 20-dim fit we find degeneracy in δC_{10}^e and $\delta C_{10}^{e\prime}$ which results in having dof$_{\rm eff}$ = 19. With the current data, there are still several of the Wilson coefficients which are only loosely constrained, especially in the electron sector where there is less

All observables with $\chi^2_{\rm SM} = 225.8$, nr. obs.= 173 **2021 fit results** ($\chi^2_{\rm min} = 151.6$; $\text{Pull}_{\rm SM} = 5.5(5.6)\sigma$)			
δC_7		δC_8	
0.05 ± 0.03		-0.70 ± 0.40	
$\delta C_7'$		$\delta C_8'$	
-0.01 ± 0.02		0.00 ± 0.80	
δC_9^μ	δC_9^e	δC_{10}^μ	δC_{10}^e
-1.16 ± 0.17	-6.70 ± 1.20	0.20 ± 0.21	degenerate w/ ↓
$\delta C_9'^\mu$	$\delta C_9'^e$	$\delta C_{10}'^\mu$	$\delta C_{10}'^e$
0.09 ± 0.34	1.90 ± 1.50	-0.12 ± 0.20	degenerate w/ ↑
$\delta C_{Q_1}^\mu$	$\delta C_{Q_1}^e$	$\delta C_{Q_2}^\mu$	$\delta C_{Q_2}^e$
0.04 ± 0.10	-1.50 ± 1.50	-0.09 ± 0.10	-4.10 ± 1.5
$\delta C_{Q_1}'^\mu$	$\delta C_{Q_1}'^e$	$\delta C_{Q_2}'^\mu$	$\delta C_{Q_2}'^e$
0.15 ± 0.10	-1.70 ± 1.20	-0.14 ± 0.11	-4.20 ± 1.2

All observables with $\chi^2_{\rm SM} = 253.5$, nr. obs.= 183 **2022 fit results** ($\chi^2_{\rm min} = 179.1$; $\text{Pull}_{\rm SM} = 5.5(5.5)\sigma$)			
δC_7		δC_8	
0.06 ± 0.03		-0.80 ± 0.40	
$\delta C_7'$		$\delta C_8'$	
-0.01 ± 0.01		-0.30 ± 1.30	
δC_9^μ	δC_9^e	δC_{10}^μ	δC_{10}^e
-1.14 ± 0.19	-6.50 ± 1.90	0.21 ± 0.20	degenerate w/ ↓
$\delta C_9'^\mu$	$\delta C_9'^e$	$\delta C_{10}'^\mu$	$\delta C_{10}'^e$
0.05 ± 0.32	1.40 ± 2.30	-0.03 ± 0.19	degenerate w/ ↑
$\delta C_{Q_1}^\mu$	$\delta C_{Q_1}^e$	$\delta C_{Q_2}^\mu$	$\delta C_{Q_2}^e$
0.04 ± 0.20	-1.60 ± 1.70	-0.15 ± 0.08	-4.10 ± 0.9
$\delta C_{Q_1}'^\mu$	$\delta C_{Q_1}'^e$	$\delta C_{Q_2}'^\mu$	$\delta C_{Q_2}'^e$
-0.03 ± 0.20	-1.50 ± 2.10	-0.16 ± 0.08	-4.00 ± 1.2

Table 4. Comparison of 20-dim fit to all observables with the 2022 (2021) result on the right (left). The $\text{Pull}_{\rm SM}$ in the parenthesis is given for $\text{dof}_{\rm eff} = 19$.

data. The significance of the NP in our 20-dim fit is 5.5σ, remaining the same as what we had found in Ref. [21].

4 Conclusions

We presented the NP fits to rare B decays which include the recent measurements of $B_s \to \phi\mu^+\mu^-$ observables and the lepton flavour violating ratios $R_{K^{*+}}$ and R_{K_S} by LHCb. We have furthermore updated the BR($B_s \to \mu^+\mu^-$) combination to include the very recent measurement by CMS. The main change in the NP fits is a reduction for the significance of a δC_{10}^μ solution or the scenarios involving it which is mainly due to the recent BR($B_s \to \mu^+\mu^-$) measurement being in agreement with the SM value. However, the hierarchy of the favoured scenarios for the global fit has remained stable and the preferred scenario is still NP with δC_9^μ. We also updated our twenty dimensional fit which avoids the look elsewhere effect finding a 5.5σ significance.

Acknowledgments

FM is grateful to the FCCP2022 organisers for their invitation to present this work.

References

[1] LHCb collaboration, *Measurement of Form-Factor-Independent Observables in the Decay $B^0 \to K^{*0}\mu^+\mu^-$*, *Phys. Rev. Lett.* **111** (2013) 191801 [1308.1707].

[2] LHCb collaboration, *Angular analysis of the $B^0 \to K^{*0}\mu^+\mu^-$ decay using 3 fb^{-1} of integrated luminosity*, *JHEP* **02** (2016) 104 [1512.04442].

[3] LHCb collaboration, *Measurement of CP-Averaged Observables in the $B^0 \to K^{*0}\mu^+\mu^-$ Decay*, *Phys. Rev. Lett.* **125** (2020) 011802 [2003.04831].

[4] T. Hurth, F. Mahmoudi and S. Neshatpour, *Model independent analysis of the angular observables in $B^0 \to K^{*0}\mu^+\mu^-$ and $B^+ \to K^{*+}\mu^+\mu^-$*, *Phys. Rev. D* **103** (2021) 095020 [2012.12207].

[5] LHCb collaboration, *Angular Analysis of the $B^+ \to K^{*+}\mu^+\mu^-$ Decay*, *Phys. Rev. Lett.* **126** (2021) 161802 [2012.13241].

[6] LHCb collaboration, *Angular analysis and differential branching fraction of the decay $B_s^0 \to \phi\mu^+\mu^-$*, *JHEP* **09** (2015) 179 [1506.08777].

[7] LHCb collaboration, *Angular analysis of the rare decay $B_s^0 \to \phi\mu^+\mu^-$*, *JHEP* **11** (2021) 043 [2107.13428].

[8] LHCb collaboration, *Branching Fraction Measurements of the Rare $B_s^0 \to \phi\mu^+\mu^-$ and $B_s^0 \to f_2'(1525)\mu^+\mu^-$ - Decays*, *Phys. Rev. Lett.* **127** (2021) 151801 [2105.14007].

[9] LHCb collaboration, *Differential branching fractions and isospin asymmetries of $B \to K^{(*)}\mu^+\mu^-$ decays*, *JHEP* **06** (2014) 133

[1403.8044].

[10] LHCb collaboration, *Differential branching fraction and angular analysis of $\Lambda_b^0 \to \Lambda\mu^+\mu^-$ decays*, *JHEP* **06** (2015) 115 [1503.07138].

[11] C. Bobeth, M. Chrzaszcz, D. van Dyk and J. Virto, *Long-distance effects in $B \to K^*\ell\ell$ from analyticity*, *Eur. Phys. J. C* **78** (2018) 451 [1707.07305].

[12] N. Gubernari, D. van Dyk and J. Virto, *Non-local matrix elements in $B_{(s)} \to \{K^{(*)}, \phi\}\ell^+\ell^-$*, *JHEP* **02** (2021) 088 [2011.09813].

[13] N. Gubernari, M. Reboud, D. van Dyk and J. Virto, *Improved Theory Predictions and Global Analysis of Exclusive $b \to s\mu^+\mu^-$ Processes*, 2206.03797.

[14] G. Hiller and F. Kruger, *More model-independent analysis of $b \to s$ processes*, *Phys. Rev. D* **69** (2004) 074020 [hep-ph/0310219].

[15] LHCb collaboration, *Test of lepton universality using $B^+ \to K^+\ell^+\ell^-$ decays*, *Phys. Rev. Lett.* **113** (2014) 151601 [1406.6482].

[16] LHCb collaboration, *Search for lepton-universality violation in $B^+ \to K^+\ell^+\ell^-$ decays*, *Phys. Rev. Lett.* **122** (2019) 191801 [1903.09252].

[17] LHCb collaboration, *Test of lepton universality in beauty-quark decays*, *Nature Phys.* **18** (2022) 277 [2103.11769].

[18] LHCb collaboration, *Test of lepton universality with $B^0 \to K^{*0}\ell^+\ell^-$ decays*, *JHEP* **08** (2017) 055 [1705.05802].

[19] LHCb collaboration, *Tests of lepton universality using $B^0 \to K_S^0\ell^+\ell^-$ and $B^+ \to K^{*+}\ell^+\ell^-$ decays*, *Phys. Rev. Lett.* **128** (2022) 191802 [2110.09501].

[20] G. D'Ambrosio, A.M. Iyer, F. Mahmoudi and S. Neshatpour, *Anatomy of kaon decays and prospects for lepton flavour universality violation*, *JHEP* **09** (2022) 148 [2206.14748].

[21] T. Hurth, F. Mahmoudi, D.M. Santos and S. Neshatpour, *More Indications for Lepton Nonuniversality in $b \to s\ell^+\ell^-$*, *Phys. Lett. B* **824** (2022) 136838 [2104.10058].

[22] M. Algueró, B. Capdevila, S. Descotes-Genon, J. Matias and M. Novoa-Brunet, *$b \to s\ell^+\ell^-$ global fits after R_{K_S} and $R_{K^{*+}}$*, *Eur. Phys. J. C* **82** (2022) 326 [2104.08921].

[23] W. Altmannshofer and P. Stangl, *New physics in rare B decays after Moriond 2021*, *Eur. Phys. J. C* **81** (2021) 952 [2103.13370].

[24] M. Ciuchini, M. Fedele, E. Franco, A. Paul, L. Silvestrini and M. Valli, *Lessons from the $B^{0,+} \to K^{*0,+}\mu^+\mu^-$ angular analyses*, *Phys. Rev. D* **103** (2021) 015030 [2011.01212].

[25] L.-S. Geng, B. Grinstein, S. Jäger, S.-Y. Li, J. Martin Camalich and R.-X. Shi, *Implications of new evidence for lepton-universality violation in $b \to s\ell^+\ell^-$ decays*, *Phys. Rev. D* **104** (2021) 035029 [2103.12738].

[26] F. Mahmoudi, *Theoretical Review of Rare B Decays*, in *20th Conference on Flavor Physics and CP Violation* , 8, 2022 [2208.05755].

[27] ATLAS collaboration, *Study of the rare decays of B_s^0 and B^0 mesons into muon pairs using data collected during 2015 and 2016 with the ATLAS detector*, *JHEP* **04** (2019) 098 [1812.03017].

[28] CMS collaboration, *Measurement of properties of $B_s^0 \to \mu^+\mu^-$ decays and search for $B^0 \to \mu^+\mu^-$ with the CMS experiment*, *JHEP* **04** (2020) 188 [1910.12127].

[29] LHCb collaboration, *Measurement of the $B_s^0 \to \mu^+\mu^-$ decay properties and search for the $B^0 \to \mu^+\mu^-$ and $B_s^0 \to \mu^+\mu^-\gamma$ decays*, *Phys. Rev. D* **105** (2022) 012010 [2108.09283].

[30] LHCb collaboration, *Analysis of Neutral B-Meson Decays into Two Muons*, *Phys. Rev. Lett.* **128** (2022) 041801 [2108.09284].

[31] CMS collaboration, *Measurement of $B_s^0 \to \mu^+\mu^-$ decay properties and search for the $B^0 \to \mu\mu$ decay in proton-proton collisions at $\sqrt{s}$ = 13 TeV*, CMS-PAS-BPH-21-006.

[32] T. Hurth, F. Mahmoudi, D. Martinez Santos and S. Neshatpour, *Lepton nonuniversality in exclusive $b \to s\ell\ell$ decays*, *Phys. Rev. D* **96** (2017) 095034 [1705.06274].

[33] A. Arbey, T. Hurth, F. Mahmoudi, D.M. Santos and S. Neshatpour, *Update on the $b \to s$ anomalies*, *Phys. Rev. D* **100** (2019) 015045 [1904.08399].

[34] F. Mahmoudi, *SuperIso: A Program for calculating the isospin asymmetry of $B \to K^*\gamma$ in the MSSM*, *Comput. Phys. Commun.* **178** (2008) 745 [0710.2067].

[35] F. Mahmoudi, *SuperIso v2.3: A Program for calculating flavor physics observables in Supersymmetry*, *Comput. Phys. Commun.* **180** (2009) 1579 [0808.3144].

[36] F. Mahmoudi, *SuperIso v3.0, flavor physics observables calculations: Extension to NMSSM*, *Comput. Phys. Commun.* **180** (2009) 1718.

[37] S. Neshatpour and F. Mahmoudi, *Flavour Physics with SuperIso*, *PoS* **TOOLS2020** (2021) 036 [2105.03428].

[38] S. Neshatpour and F. Mahmoudi, *Flavour Physics Phenomenology with SuperIso*, *PoS* **CompTools2021** (2022) 010 [2207.04956].

[39] BELLE collaboration, *Test of lepton flavor universality and search for lepton flavor violation in $B \to K\ell\ell$ decays*, *JHEP* **03** (2021) 105 [1908.01848].

[40] T. Hurth, F. Mahmoudi and S. Neshatpour, *Implications of the new LHCb angular analysis of $B \to K^*\mu^+\mu^-$: Hadronic effects or new physics?*, *Phys. Rev. D* **102** (2020) 055001 [2006.04213].

[41] V.G. Chobanova, T. Hurth, F. Mahmoudi, D. Martinez Santos and S. Neshatpour, *Large hadronic power corrections or new physics in the rare decay $B \to K^*\mu^+\mu^-$?*, *JHEP* **07** (2017) 025 [1702.02234].

[42] S. Neshatpour, V.G. Chobanova, T. Hurth, F. Mahmoudi and D. Martinez Santos, *Direct comparison of global fits to the $B \to K^*\mu^+\mu^-$ data assuming hadronic corrections or new physics*, in *52nd Rencontres de Moriond on QCD and High Energy Interactions*, pp. 87–90, 2017 [1705.10730].

[43] S. Neshatpour, T. Hurth and F. Mahmoudi, *Searching for New Physics with $B^0 \to K^{*0}\mu^+\mu^-$*, *PoS* **ICHEP2020** (2021), 406 [2012.07408].

[44] CMS collaboration, *Angular analysis of the decay $B^+ \to K^+\mu^+\mu^-$ in proton-proton collisions at $\sqrt{s} = 8$ TeV*, *Phys. Rev. D* **98** (2018) 112011 [1806.00636].

[45] LHCb collaboration, *Strong constraints on the $b \to s\gamma$ photon polarisation from $B^0 \to K^{*0}e^+e^-$ decays*, *JHEP* **12** (2020) 081 [2010.06011].

[46] T. Hurth, F. Mahmoudi and S. Neshatpour, *On the anomalies in the latest LHCb data*, *Nucl. Phys. B* **909** (2016) 737 [1603.00865].

[47] T. Hurth, F. Mahmoudi and S. Neshatpour, *Global fits to $b \to s\ell\ell$ data and signs for lepton non-universality*, *JHEP* **12** (2014) 053 [1410.4545].

[48] A. Arbey, T. Hurth, F. Mahmoudi and S. Neshatpour, *Hadronic and New Physics Contributions to $b \to s$ Transitions*, *Phys. Rev. D* **98** (2018) 095027 [1806.02791].

https://doi.org/10.1051/epjconf/202328901003

Precision Measurement of the W Mass

Giorgio Chiarelli[1]*, *for the CDF Collaboration*

[1]Istituto Nazionale di Fisica Nucleare, Sezione di Pisa, 56127 Pisa, Italy

Abstract. Using proton-antiproton collision data corresponding to an integrated luminosity of 8.8 fb^{-1} collected in Tevatron Run 2, CDF performed a high precision measurement of the W mass. This result, 80433.5±6.4±6.9 MeV, is the most precise so far, and shows a discrepancy of 7σ with Standard Model expectations.

1 Introduction

Precision measurement of the W mass is a key ingredient in understanding the standard model of fundamental interactions. At tree level it is linked to the Z mass via the Weinberg angle ϑ_W ($\cos\vartheta_W = M_W/M_Z$), while through radiative corrections is linked to the top mass and opens a window on new physics.

Table 1. Events used in this analysis.

Sample	N. of events
W→eν	1811799
W→eν	2424486
Z→ee	66180
Z→μμ	238534

The previous CDF result was based on 2.2fb^{-1} of data, with an uncertainty (19 MeV) evenly due to statistical and systematic uncertainty. At the end of Run 2, we had a four-fold set of proton-antiproton collisions on tape. The large data yield available spurred an analysis strategy designed for reducing systematics uncertainty at or below that due to statistics.

2. Analysis

At the Tevatron most (~80%) of the contribution to W production was due to valence quark-antiquark annihilation, and the W boson decays either hadronically to a quark pair, or leptonically to a charged lepton plus a neutrino. In this analysis we use events in which W decays into and electron (or muon) and a neutrino. Since the transverse momentum of the lepton carries most of the information on the W mass, precision measurement of the lepton kinematics is the key ingredient for this study.

2.1 W and Z samples

The sample was collected by a set of inclusive lepton triggers requiring a high P_T charged lepton (electron or muon). At offline level, a selection requiring either a track (P_T>18 GeV[†]) matching an electromagnetic calorimetric cluster (E_T>30 GeV), or a high P_T (>30 GeV) track and loose lepton ID requirements, provides the first data set. W candidates are required to have a neutrino with P_T>30 GeV, $|u_T|$<15[‡] GeV, and transverse mass 60<M_T<100 GeV. Events with a second charged leptons are selected as Z candidates. Both **μ** and ***e*** data come from the same 8.8fb^{-1} sample, and the event selection provide a clean (mis-id <0.5%) data set (Table 1). The W/Z ratio scales with geometric acceptance of the CDF detector [1].

2.2 Analysis strategy

Three main goals set the overall analysis strategy:

- Robustness. Obtained by constraining the same relevant parameter in as many ways as possible.
- Precision. Combine independent, yet consistent measurements.
- Minimize bias. Perform blind analyses for the *Z* and *W* samples.

In the final fit we combine three observables (P_T of the charged and neutral leptons, M_T of the W) for each lepton flavour, for a total of six.

Signal generation and simulation of the detector response is a basic ingredient. Our Monte Carlo is custom-made; therefore, all tracking effects are based on

* Corresponding author: giorgio.chiarelli@pi.infn.it
† Through this paper *c* will be set to 1.
‡ u_T is the transverse momentum of hadronic recoil, computed as the (2-dimensional) vector sum of the energy recorded in the individual calorimetric towers.

first principles (from ionization losses to non-gaussian tail of multiple scattering). In this way detector response can be directly checked against data. Particles are propagated through the material, simulated using a 3D-look-up table that starts from a database of construction data, then tuned using conversion electrons.

2.3 Addressing Systematics

Generator level input for W and Z boson is provided by RESBOS. QED effects are generated with PHOTOS and checked against HORACE[§]; after detailed studies and many cross-checks, the overall uncertainty on the W mass due to QED corrections is tiny (3 MeV). We also use our data to tune RESBOS, by fitting the $P_T(ll)$ distribution in Z events, therefore reducing its contribution to M_W uncertainty to 1.8 MeV. Our modelling is checked by comparing the angular variable ϕ^*_η ** in data and MC [1].

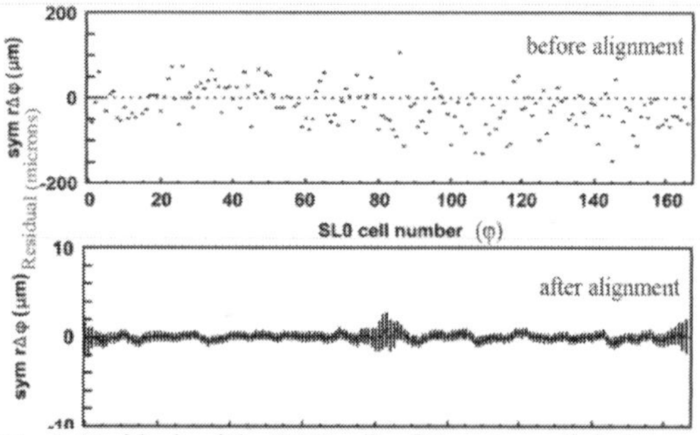

Fig. 1. Residuals of the COT cells of SuperLayer 0, before and after alignment procedure. The vertical scale reduces from 200 to 10 μm.

The most important ingredients of the W mass are the energy scale and calibration of the detectors used. The two most important ones are the Central Outer tracker (COT) and the EM calorimeter. CDF performed an individual calibration of the 2,520 cells of the COT using cosmic rays [2]. The excellent result of this procedure can be appreciated by looking at Fig. 1 where residuals of cells belonging to SuperLayer 0 of the COT are shown before (scale ±200 μm) and after (scale ±10 μm) alignment. The accurate procedure with cosmic rays removes most deformations. Remaining effects are addressed by looking at <E/p> for electrons and positrons.

The tracking momentum scale was obtained by using J/Ψ→μμ, and Υ→ μμ data. As Υ are produced promptly, the latter data set allows a check of beam constrained and beam unconstrained fitting procedures. In this analysis we resolved a previous discrepancy between the two procedures, therefore removing it.

The Z mass reconstructed in the dimuon channel is key to set the final energy scale. In Fig. 2 we show the Δp/p in bins of $1/P_T$ for the various data sample (J/Ψ, Υ, Z). In a first step we combine the J/Ψ and Υ samples, obtaining a correction (-1393±26) ppm.

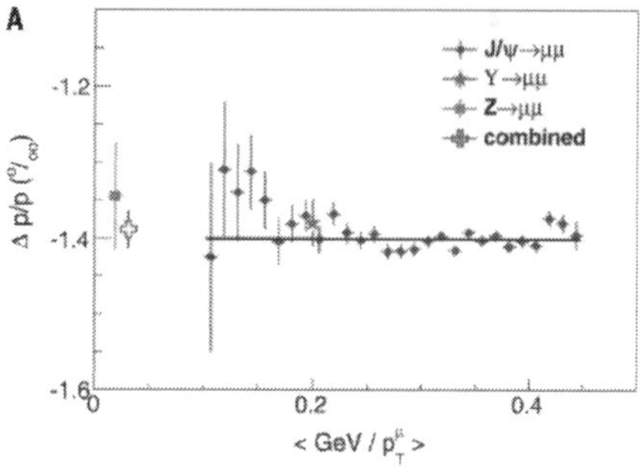

Fig. 2. Correction to momentum scale as a function of P_T of the muon in different data sample.

We use this value, and, after unblinding, we measure the Z mass: $91192 \pm 6.4_{stat} \pm 4.0_{syst}$ MeV, consistent with WA (91187.6±2.1 MeV). This result is also the most precise Z mass determination at an hadron collider. We combine this measurement with results from J/Ψ and Υ, and obtain a final correction Δp/p = (-1389±25) ppm, as shown (red open cross) in Fig. 2. The final uncertainty on M_W due to tracking scale is 2 MeV.

The EM calorimeter is the second detector playing a major role in this measurement. We use an energy response derived from a detailed GEANT4 simulation tuned on data to reproduce electromagnetic shower features (e. g. energy leakage, angular dependence, non-linearity). The energy scale is set by using the peak of the E/p spectrum. The high-end tail of the same distribution provides information on the radiative material, and the low-tail side on energy leakage. We use this additional knowledge to tune the material map in our simulation: as an example, the E/p tail shows an excellent data-MC agreement after scaling up X_0 by 1.049±0.002. A small non-linear response of the EM calorimeter is tuned using W and Z data.

Table 2. Z mass in different samples using calorimetric clusters or reconstructed tracks, results in MeV.

e+/e- sample	Calorimeter	Tracking
E/p<1	91190.9±19.7	91215.2±22.4
E/p>1 &E/p<1	91201.1±21.5	91259.9±39.0
E/p>1 only	91184.5±46.4	91167.7±109.9

3. Cross Checks

As both Z and W data are linked through the energy scale, the actual value of their masses is blinded through all the analysis. After unblinding, using the energy scale determined from a fit to the E/p distribution we fit the Z mass. The result is consistent within 0.5σ

§ In collaboration with C.M. Carloni Calame, G. Montagna, O. Nicrosini, A.Vicini, that we would like to thank for the help.

** $\phi^*_\eta = \tan((\pi\text{-}\Delta_{ll})/2)\mathrm{sech}((\eta^- - \eta^+)/2)$

with PDG, therefore, we combine our Z mass with E/p for the best accuracy in the energy scale determination. We also study our understanding of tracking by determining the Z mass reconstructed using electron track information. Several cross checks can be performed, for example by measuring mass reconstructed using e^+e^- affected by bremsstrahlung or belonging to different h regions. A sample of results is shown in Table 2, where the good agreement among the various samples (in this case separated in E/p regions)

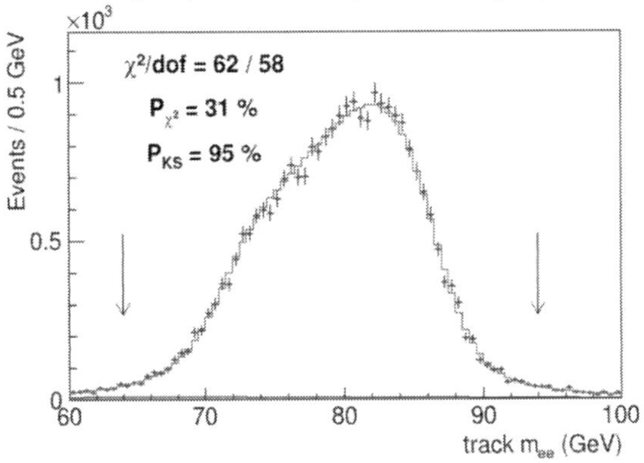

Fig. 3. Mass distribution in Z→ee events from the electron tracks.

can be appreciated. In Fig. 3 we show the dielectron mass distribution obtained by the tracking alone.

Z→ll data is also key to tune the detector response to hadronic recoil u_T. In this instance we exploit the similarity between W and Z events topologies and study the P_T imbalance in Z events.

3.1 Further constraints from data

Besides $P_T(Z)$, in this analysis we use P_T (W)/ P_T (Z) to constrain the boson P_T spectrum. A triple differential cross section, computed at NNLO, is used to model the scale variation of the ratio, and we also consider the correlation with the hadronic recoil.

Syst.	This analysis	2.2 fb^{-1}
PDFs	3.9	10
Background	3.3	3
Lepton energy scale	3	7
QED radiation	2.7	4
P_TZ model	1.8	5 [P_T(W)]
P_TW/P_TZ	1.3	

Table 3. Major systematics in new analysis compared to previous one. Values in MeV.

For the Parton Distribution Functions (PDFs) we use NNPDF3.1 as the default NNLO PDFs, and we check against several PDFs: NNPDF3.1, CT18 and MMHT2014 (NNLO) agree within ±2.1 MeV. A similar comparison with PDFs at NLO finds that results are consistent within ±3 MeV. The final uncertainty of 3.9 MeV due to the PDFs is estimated by varying 25 symmetric eigenvectors built following the procedure described in [3].

In Table 3 we report a summary of the most relevant systematics compared to previous analysis. After the publication of this measurement, RESBOS authors compared our findings, based on RESBOS1, and constrained using data, with newer RESBOS2 Monte Carlo. They find agreement with our results [4].

Table 4. Fit results of W boson mass. All figures in MeV, first uncertainty statistical, second systematic.

	Muon chann.	Electron chann.
$m_T(l,\nu)$	80446.1±9.2±7.3	80429.1±10.3±8.5
$P_T(l)$	80428.2±9.6±10.3	80411.4±10.7±11.8
$P_T(\nu)$	80428.9±13.1±10.9	80426.3±14.5±11.7
Comb.	80433.5±6.4±6.9	

4. Results

The P_Tl, $P_T\nu$, and $M_T(W)$ distributions in ***e*** and μ samples are fit simultaneously. The shaded regions in Fig. 4 represent the background contributions in the different channels. All backgrounds are below 1% except for the Z→μμ channel (7.37±0.1%) where one of the two muons is lost due to geometric acceptance.

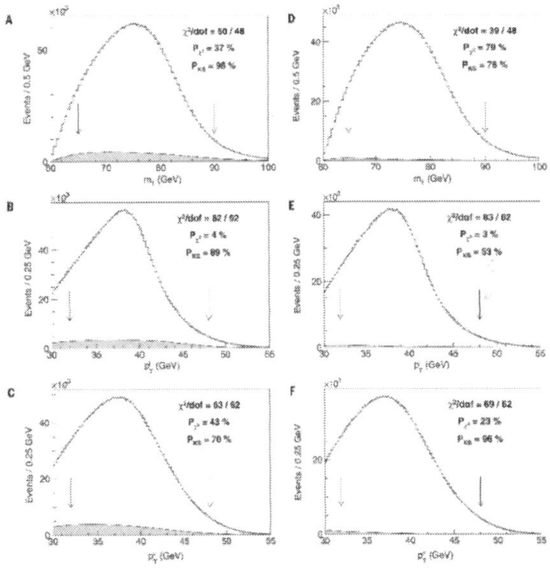

Fig. 4. Fits of data (blue points) and simulated samples (red histograms) used to obtain the W boson mass. Arrows indicate the range used in the fit. Shade regions indicate the backgrounds. Left row: muon channel; top to bottom: M_T, P_T charged lepton, P_T neutrino.

The procedure is blinded: M_Z and M_W are assigned an (unknown) offset in the [-50, +50] MeV range around their values, so that effects can be studied without biasing the result. Once the analysis was completed and thoroughly reviewed, we removed the offset and obtained the final value, M_W=80433.5±6.4±6.9 MeV. In

Fig. 4 we show comparison of data overlayed to Monte Carlo simulated data, and Table 4 reports the fit results for different channels and variables. The final, combined, value disagrees with expectations from SM global fit of $80357 \pm 4_{input} \pm 4_{th}$ MeV [5] (see Fig. 4).

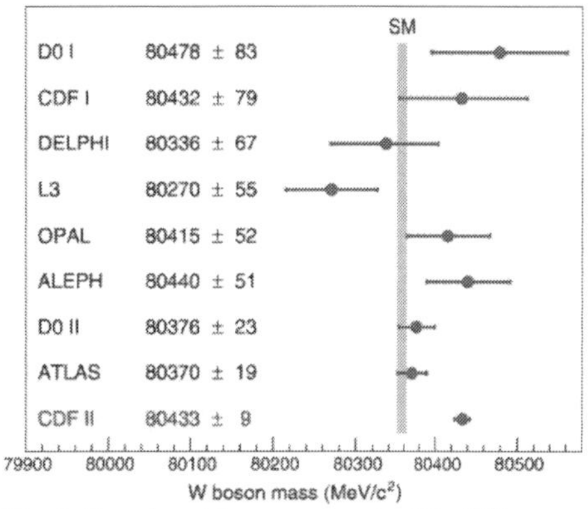

Fig. 5. Experimental measurements of the W boson mass.

The list of improvements with respect to the previous measurement is reported in Table 5. The first two contributions are additive and shift the 2.2fb^{-1} result by +13.5 MeV.

Table 5. Improvements with respect to 2.2fb^{-1} analysis.

Method or technique	impact
Detailed treatment of parton distribution functions	+3.5 MeV
Resolved beam-constraining bias in CDF reconstruction	+10 MeV
Improved COT alignment and drift model	uniformity
Improved modeling of calorimeter tower resolution	uniformity
Temporal uniformity calibration of CEM towers	uniformity
Lepton removal procedure corrected for luminosity	uniformity
Higher-order calculation of QED radiation in J/ψ and Y decays	accuracy
Modeling kurtosis of hadronic recoil energy resolution	accuracy
Improved modeling of hadronic recoil angular resolution	accuracy
Modeling dijet contribution to recoil resolution	accuracy
Explicit luminosity matching of pileup	accuracy
Modeling kurtosis of pileup resolution	accuracy
Theory model of p^W/p^Z spectrum ratio	accuracy
Constraint from p^W data spectrum	robustness
Cross-check of p^Z tuning	robustness

4. Conclusion

Using 8.8 fb^{-1} of data collected at the Tevatron Run 2, CDF performed a high precision measurement of the W mass: 80433.5±9.4 MeV. Thanks to improvements in our analysis, as well on the theoretical aspects relevant to this measurement, our new result is a factor of two more accurate than the old one. A summary of experimental results is shown in Fig. 5. The deviation from SM prediction of 80357±6 MeV shows a significance of 7σ, suggesting the possibility of improvements in SM calculations or an extension of the SM itself. As an example, in Fig. 6, we report this measurement along with earlier LEP2/Tevatron average and top quark mass WA [5] compared to theoretical prediction [6].

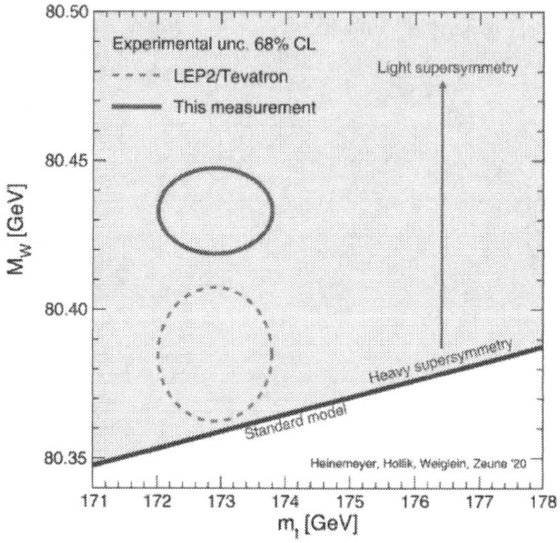

Fig. 6. W boson and top quark mass measurements and theoretical predictions.

References

1. T. Aaltonen et al. CDF Collaboration, Science 376, 6589 (2022), https://doi.org/10.1126/science.abk1781.
2. A.V. Kotwal and C.P. Hays, Nucl. Instr. Meth. A 762 (2014).
3. S. Carrazza, S. Forte, Z. Kassabov, J. I. Latorre and J. Rojo, Eur. Phys. J. C 75, 369 (2015).
4. J. Isaacson, Y. Fu, C.-P. Yuan, arXiv:2205.02788.
5. P.A. Zyla et al. (Particle Data Group), Prog. Theor. Exp. Phys. 2020, 083C01 (2020).
6. S. Heinemeyer, W. Hollik, G. Weiglein, L. Zeune, J. High Energy Phys. 2013, 84 (2013).

Dispersive determination of the HVP contribution to the muon $g-2$

Martin Hoferichter[1,*]

[1] Albert Einstein Center for Fundamental Physics, Institute for Theoretical Physics, University of Bern, Sidlerstrasse 5, 3012 Bern, Switzerland

Abstract. The determination of hadronic vacuum polarization (HVP) from $e^+e^- \to$ hadrons cross-section data, in the energy region relevant for the anomalous magnetic moment of the muon, has recently been challenged by lattice-QCD calculations, especially for the intermediate window in Euclidean time. In these proceedings we review some frequently-asked questions on the comparison between data-driven and lattice-QCD evaluations of the HVP contribution.

1 Introduction

Traditionally, the HVP contribution to a_μ has been ev: ated using cross-section data for $e^+e^- \to$ hadrons, lead to the current 4.2σ tension between Standard-Model ļ diction [1–25] and the experimental world average [: 30]. However, this data-driven evaluation lies 2.1σ lo than the lattice-QCD calculation by BMWc [31], wh was subsequently confirmed by other collaborations least for the intermediate window quantity (following definition with parameters from Ref. [32]). For this ļ tial quantity, the current situation is summarized in Fig indicating a much stronger tension than observed glob: by BMWc.[1]

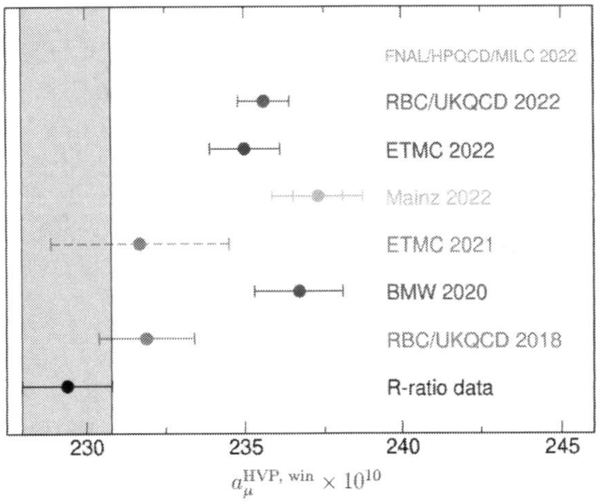

Figure 1. Intermediate window from RBC/UKQCD 2018 [32], BMW 2020 [31], ETMC 2021 [33], Mainz 2022 [34], ETMC 2022 [35], and RBC/UKQCD 2022 [36]. Note that ETMC 2021, only published as proceedings, is superseded by ETMC 2022, while RBC/UKQCD 2022 supersedes RBC/UKQCD 2018. The R-ratio band is taken from Ref. [37]. FNAL/HPQCD/MILC 2022 [38] found agreement with other collaborations for the light-quark-connected correlator (as have Refs. [39–41]), with the subleading corrections required for the full $a_\mu^{\text{HVP, win}}$ forthcoming.

The consequences of this pattern have been studie the literature to the extent that this is possible based the currently available information, including the relat... to the global electroweak fit via the hadronic running of the fine-structure constant [55–58] and changes to the 2π cross section [59]. In particular, the latter analysis shows that while, in principle, modifications to the 2π cross section consistent with analyticity and unitarity constraints would be able to accommodate the global difference to the BMWc result, a simultaneous explanation of the tension in the intermediate window is not possible. Assuming a rather uniform shift in the low-energy 2π region as well as no significant negative shifts, the form of the kernel functions suggests that at least 40% of the changes would need to originate from the energy region above 1 GeV. Such a conclusion is line with direct lattice-QCD results for the hadronic running itself [31, 60], given that the moderate increase observed therein implies that the changes cannot occur too high in energy, because then the effect on the running would be too large, but also not concentrated at very low energies, as in this case no effect would be visible at all.

Extracting more detailed information on the energy dependence is a complicated endeavor, amounting to the inversion of a Laplace transform, but could be possible if finer windows become available [37], with linear combinations to be optimized according to lattice and data-driven covariance matrices, or by considering spectral-

*e-mail: hoferichter@itp.unibe.ch

[1]For hadronic light-by-light scattering there is good agreement between lattice-QCD and phenomenological evaluations, with recent work [42–53] aimed at improving the precision by another factor 2 to match the target anticipated for the final result from the Fermilab experiment [54].

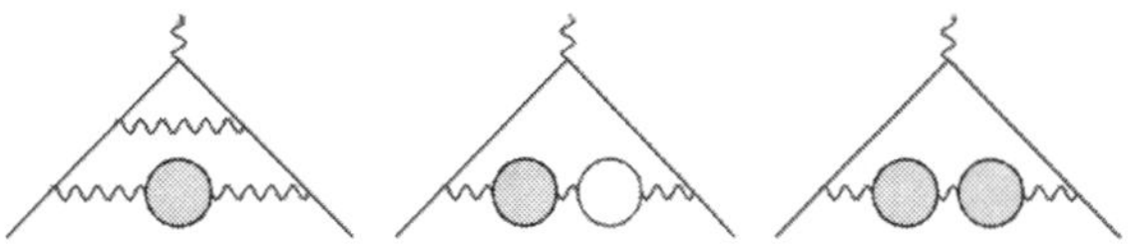

Figure 2. Higher-order insertions of HVP (gray blob). Diagram (a) includes additional photon corrections as well as muon VP, diagram (b) VP from $\ell = e, \tau$, and diagram (c) two HVP insertions.

Figure 3. Critical diagram for defining QED corrections in lattice QCD (solid lines refer to quarks, the wiggly line to the photon, and gluons are implied everywhere), diagram (f) F in the notation of Ref. [32]. The one-particle-reducible piece without any additional gluons needs to be subtracted.

weight sum rules [61]. At the moment, the situation summarized in Fig. 1 presents a puzzle, whose resolution remains far from obvious. It is therefore prudent to reconsider the relation between lattice-QCD calculations and data-driven evaluations in more detail, here, we review some of the frequently-asked questions that arise in this context.

2 Conventions for higher-order corrections

By convention, the cross section that enters the HVP master formula [62, 63]

$$a_\mu^{\text{HVP, LO}} = \left(\frac{\alpha m_\mu}{3\pi}\right)^2 \int_{s_{\text{th}}}^\infty ds \frac{\hat{K}(s)}{s^2} R_{\text{had}}(s),$$
$$R_{\text{had}}(s) = \frac{3s}{4\pi\alpha^2}\sigma(e^+e^- \to \text{hadrons}(+\gamma)), \tag{1}$$

is defined in a photon-inclusive way, including radiative intermediate states and final-state radiation (FSR). Initial-state radiation (ISR) and VP are subtracted to avoid double counting with higher-order HVP insertions. This bookkeeping is illustrated in Fig. 2, where diagram (a) gives photonic corrections and muon VP, diagram (b) VP corrections from $\ell = e, \tau$ (leading to a dependence on the respective lepton mass ratio), and diagram (c) the double HVP iteration. Numerically, these three classes sum up to [11, 64–66]

$$a_\mu^{\text{HVP, NLO}} \simeq [-20.7 + 10.6 + 0.3] \times 10^{-10}$$
$$= -9.8 \times 10^{-10}, \tag{2}$$

in such a way that the double HVP iteration, which potentially requires some care in the definition of the hadronic two-point function, only produces a very small effect. Defining each VP function as the one-particle-irreducible amplitude, in accordance with the running of the fine-structure constant

$$\alpha(q^2) = \frac{\alpha(0)}{1 - \Delta\alpha_{\text{lep}}(q^2) - \Delta\alpha_{\text{had}}(q^2)},$$
$$\Delta\alpha_{\text{had}}(q^2) = -\frac{\alpha q^2}{3\pi} P \int_{s_{\text{th}}}^\infty ds \frac{R_{\text{had}}(s)}{s(s - q^2)}, \tag{3}$$

one can thus unfold the experimental cross sections and iteratively determine the R-ratio $R_{\text{had}}(s)$ to be inserted in Eq. (1). In this way, it suffices to consider the exclusive hadronic channels that contribute, without the need to ever specify the role of a specific resonance such as ρ, ω, ϕ. The only subtlety concerns very narrow $c\bar{c}$ and $b\bar{b}$ resonances for which the Dyson series no longer converges [7, 67]. In these cases, one needs to take out the resonance that is being corrected in R_{had} in the VP undressing, but such a procedure is not even necessary for the low-energy cross sections that dominate the HVP integral. The same conventions are used also for yet higher-order HVP insertions [13, 68].

To match these conventions in lattice QCD, isospin-breaking (IB) corrections need to be included, both arising from QED and strong IB. The latter is, in principle, straightforward to include by means of insertions of the quark mass difference, while the QED corrections need to be defined in accordance with the conventions for $a_\mu^{\text{HVP, LO}}$ in the data-driven approach [32]. This mainly concerns the quark-level diagram shown in Fig. 3, which only becomes one-particle-irreducible if dressed with additional gluon interactions between the quark lines. Accordingly, the diagram without any additional gluons needs to be subtracted, as can be implemented by subtracting the separate quantum averages of the quark loops. The final sum of isospin-limit calculation and IB corrections can then be directly compared to phenomenology, and it is this comparison that is shown in Fig. 1.[2]

The size of the IB corrections can also be estimated directly from phenomenology, either by summing up contributions from the exclusive channels [69] or, in the case of strong IB, from chiral perturbation theory [70], with the critical low-energy constant $\delta C_{93}^{(1)}$ extracted from τ decays. In the exclusive approach, sizable effects arise from the radiative channels $\pi^0\gamma$, $\eta\gamma$, FSR, ρ–ω mixing, and pion-mass effects in 2π [71, 72], as well as similar, resonance-enhanced effects in $\bar{K}K$ [73]. Separating all effects into QED and strong contributions, and decomposing onto the Euclidean-time windows, there is reasonable agreement with Refs. [31, 32], especially in view of limitations in the phenomenological approach due to unknown IB in resonance couplings and the truncation in the sum over hadronic channels. In fact, Refs. [69, 70] both indicate a larger strong IB correction for the full HVP contribution than Ref. [31] (albeit largely consistent within uncertainties) and Ref. [69] finds an increased QED correction in

[2]Note that at present only Refs. [31, 32] provide complete calculations of the IB corrections, while Ref. [34] uses a partial estimate and Ref. [35] the result from Ref. [31]. Both QED and strong IB corrections are found to be small for the intermediate window.

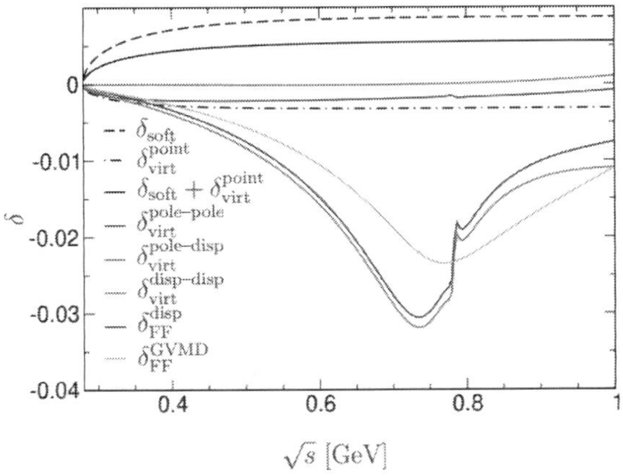

Figure 4. Correction factor δ that determines the forward–backward asymmetry in $e^+e^- \to \pi^+\pi^-$ for scattering angle $z = \cos(1)$. The black lines refer to the point-like, scalar-QED result (dashed/dot-dashed/solid to real/virtual/sum). The structure-dependent corrections are separated into pole–pole (blue), pole–dispersive (red), and dispersive–dispersive (green) contributions. Their sum minus the point-like result gives the maroon curve, which can be compared to the GVMD model calculation from Ref. [89] (orange). Figure taken from Ref. [90].

the intermediate window. In either case the lattice-QCD result for HVP would increase even further if these phenomenological estimates were adopted, suggesting that issues with IB corrections are unlikely to resolve the observed tension.

3 Radiative corrections and Monte-Carlo generators

In practice, a key question becomes whether radiative corrections as correspond to the conventions in Sec. 2 can be controlled and implemented in Monte-Carlo (MC) generators at the sufficient level of precision. The status of MC generators around the time the BaBar [74, 75] and KLOE [76–79] 2π data appeared is documented in detail in Ref. [80]. From a theory perspective the formalism is based on scalar QED supplemented by the pion vector form factor whenever possible ("FsQED"), based on the observation that this strategy should capture the dominant, infrared-enhanced (IR) effects [81–84]. Corrections beyond IR enhancement were evaluated in Ref. [85] for the $\pi\pi\gamma$ channel and found to be small. Another possible concern regarding the ISR data sets arose from the fact that the MC generator PHOKHARA [86, 87] did not include full NLO corrections, but these previously only estimated diagrams were recently incorporated, not revealing any sizable corrections [88].

Going forward, improved MC generators are key for the success of future experimental campaigns [91]. Recent developments involve the forward–backward asymmetry in $e^+e^- \to \pi^+\pi^-$, which can be considered a valuable test case, since due to the C-odd nature the asymmetry vanishes at tree level, while radiative corrections generate a non-vanishing result from the interference of ISR and FSR contributions [92–94]. In Ref. [89] it was observed that scalar QED without form-factor modifications was not able to describe preliminary CMD-3 data on the asymmetry, but from the perspective of dispersion relations the required form-factor corrections do correspond to the IR enhanced effects, as expected [90]. Ultimately, this is the reason why the full dispersive calculation largely reproduces the outcome of the model estimate, see Fig. 4, as the dominant pole–dispersive contribution receives both significant IR and form-factor enhancements. The study of the C-even case is in progress [95], and it remains to be seen if the dispersive analysis reveals unexpectedly large effects. The test case of the asymmetry, however, makes it appear unlikely that such effects can resolve the tension with lattice QCD either.

4 τ data

The use of τ data for the evaluation of the HVP contribution relies on the relation between the vector-current-mediated decay $\tau^- \to X^-\nu_\tau$ and the cross section $e^+e^- \to X^0$ involving the corresponding isovector final state X^0, which becomes exact in the limit of isospin invariance [96]. In practice, this implies that IB corrections need to be controlled very accurately if data for $X = 2\pi$ [97, 98] are to be included in the HVP evaluation. Phenomenological estimates of these corrections, including Refs. [99–102], are typically separated into FSR, phase-space factors, long-range radiative corrections, and the pion form factors corresponding to the two charge channels of the $\rho(770)$. Here, the need to argue in terms of hadronic states, with ρ resonance parameters to be determined from elsewhere, incurred a significant model dependence, which, together with the superior precision from e^+e^- data, ultimately led to τ data not being considered in most data-driven HVP evaluations [1].

Given the tension between lattice QCD and e^+e^- data, an independent cross check via τ decays would be highly valuable, also in view of potential experimental improvements at Belle II [103], but would require a convincing solution to the issue of IB corrections. Recent progress in this direction was reported using lattice QCD [104], allowing one to address the calculation of IB effects without the need to specify ρ properties. Challenges in this program include the matching to the experimental observables, e.g., the 2π channel needs to be isolated from the inclusive lattice calculation and long-range QED corrections need to be implemented in a consistent manner. For this last step, also the interplay with dispersive methods developed in the context of radiative corrections in $e^+e^- \to \pi^+\pi^-$ [95] could prove beneficial.

5 Avenues for future progress

Prospects for improved predictions of a_μ in the Standard Model are described in detail in Ref. [105]. In lattice QCD, this includes full calculations from other collaborations at a level comparable to BMWc, given that a comparison as in Fig. 1 is so far only available for the intermediate window. Moreover, results for other windows are expected, both as a cross check that their sum reproduces the full HVP integral and as a diagnostic tool to scrutinize first consistency among lattice-QCD calculations and then the comparison to e^+e^- data.

On the data-driven side, new input has already become available after Ref. [1], for 2π [106], 3π [107, 108], and inclusive measurements [109], but so far with a small impact on global HVP compilations.[3] Addressing the longstanding tension between KLOE and BaBar will require new data at a similar level of precision, as are expected from CMD-3 [110], BaBar [91], BESIII [111], and Belle II [103] in the coming years. Together with continued scrutiny of radiative corrections and, potentially, new input from τ decays, this should allow one to corroborate the situation for time-like data-driven evaluations.

In addition, the MUonE experiment [112–114] aims at a completely independent determination of the HVP contribution, by means of a space-like extraction from muon–electron scattering. Controlling the systematics of the cross section at a level of 10^{-5} presents unique challenges to both experiment and theory [115], with a resulting measurement that would give a direct determination in the space-like region and thus be fully complementary to the time-like approach.

In conclusion, the present situation in the determination of HVP remains puzzling, especially given the increased tension in the intermediate window shown in Fig. 1, to the extent that even scenarios for beyond-Standard-Model contamination in the e^+e^- data have been considered [116–118]. To resolve this puzzle, a large, concerted effort in lattice QCD, experiment, and phenomenology is currently underway, which should allow for critical new insights on the timescale of the Fermilab experiment.

Acknowledgements

I thank Gilberto Colangelo and Peter Stoffer for longstanding collaboration on many of the topics discussed in these proceedings, as well as for comments on the manuscript. Support by the Swiss National Science Foundation, under Project No. PCEFP2_181117, is gratefully acknowledged.

References

[1] T. Aoyama et al., Phys. Rept. **887**, 1 (2020), 2006.04822

[2] T. Aoyama, M. Hayakawa, T. Kinoshita, M. Nio, Phys. Rev. Lett. **109**, 111808 (2012), 1205.5370

[3] T. Aoyama, T. Kinoshita, M. Nio, Atoms **7**, 28 (2019)

[4] A. Czarnecki, W.J. Marciano, A. Vainshtein, Phys. Rev. D **67**, 073006 (2003), [Erratum: Phys. Rev. D **73**, 119901 (2006)], hep-ph/0212229

[5] C. Gnendiger, D. Stöckinger, H. Stöckinger-Kim, Phys. Rev. D **88**, 053005 (2013), 1306.5546

[6] M. Davier, A. Hoecker, B. Malaescu, Z. Zhang, Eur. Phys. J. C **77**, 827 (2017), 1706.09436

[7] A. Keshavarzi, D. Nomura, T. Teubner, Phys. Rev. D **97**, 114025 (2018), 1802.02995

[8] G. Colangelo, M. Hoferichter, P. Stoffer, JHEP **02**, 006 (2019), 1810.00007

[9] M. Hoferichter, B.L. Hoid, B. Kubis, JHEP **08**, 137 (2019), 1907.01556

[10] M. Davier, A. Hoecker, B. Malaescu, Z. Zhang, Eur. Phys. J. C **80**, 241 (2020), [Erratum: Eur. Phys. J. C **80**, 410 (2020)], 1908.00921

[11] A. Keshavarzi, D. Nomura, T. Teubner, Phys. Rev. D **101**, 014029 (2020), 1911.00367

[12] B.L. Hoid, M. Hoferichter, B. Kubis, Eur. Phys. J. C **80**, 988 (2020), 2007.12696

[13] A. Kurz, T. Liu, P. Marquard, M. Steinhauser, Phys. Lett. B **734**, 144 (2014), 1403.6400

[14] K. Melnikov, A. Vainshtein, Phys. Rev. D **70**, 113006 (2004), hep-ph/0312226

[15] P. Masjuan, P. Sánchez-Puertas, Phys. Rev. D **95**, 054026 (2017), 1701.05829

[16] G. Colangelo, M. Hoferichter, M. Procura, P. Stoffer, Phys. Rev. Lett. **118**, 232001 (2017), 1701.06554

[17] G. Colangelo, M. Hoferichter, M. Procura, P. Stoffer, JHEP **04**, 161 (2017), 1702.07347

[18] M. Hoferichter, B.L. Hoid, B. Kubis, S. Leupold, S.P. Schneider, Phys. Rev. Lett. **121**, 112002 (2018), 1805.01471

[19] M. Hoferichter, B.L. Hoid, B. Kubis, S. Leupold, S.P. Schneider, JHEP **10**, 141 (2018), 1808.04823

[20] A. Gérardin, H.B. Meyer, A. Nyffeler, Phys. Rev. D **100**, 034520 (2019), 1903.09471

[21] J. Bijnens, N. Hermansson-Truedsson, A. Rodríguez-Sánchez, Phys. Lett. B **798**, 134994 (2019), 1908.03331

[22] G. Colangelo, F. Hagelstein, M. Hoferichter, L. Laub, P. Stoffer, Phys. Rev. D **101**, 051501 (2020), 1910.11881

[23] G. Colangelo, F. Hagelstein, M. Hoferichter, L. Laub, P. Stoffer, JHEP **03**, 101 (2020), 1910.13432

[24] T. Blum et al. (RBC/UKQCD), Phys. Rev. Lett. **124**, 132002 (2020), 1911.08123

[25] G. Colangelo, M. Hoferichter, A. Nyffeler, M. Passera, P. Stoffer, Phys. Lett. B **735**, 90 (2014), 1403.7512

[26] G.W. Bennett et al. (Muon $g-2$), Phys. Rev. D **73**, 072003 (2006), hep-ex/0602035

[27] B. Abi et al. (Muon $g-2$), Phys. Rev. Lett. **126**, 141801 (2021), 2104.03281

[28] T. Albahri et al. (Muon $g-2$), Phys. Rev. A **103**, 042208 (2021), 2104.03201

[29] T. Albahri et al. (Muon $g-2$), Phys. Rev. Accel. Beams **24**, 044002 (2021), 2104.03240

[30] T. Albahri et al. (Muon $g-2$), Phys. Rev. D **103**, 072002 (2021), 2104.03247

[3]Note, however, that in Ref. [71] tensions with analyticity and unitarity constraints were pointed out for the SND data set [106], as were tensions in the 2π data base in the context of the ρ–ω interference.

[31] S. Borsanyi et al. (BMWc), Nature **593**, 51 (2021), 2002.12347
[32] T. Blum et al. (RBC/UKQCD), Phys. Rev. Lett. **121**, 022003 (2018), 1801.07224
[33] D. Giusti, S. Simula, PoS **LATTICE2021**, 189 (2022), 2111.15329
[34] M. Cè et al. (2022), 2206.06582
[35] C. Alexandrou et al. (ETM) (2022), 2206.15084
[36] C. Lehner (2022), https://indico.ph.ed.ac.uk/event/112/contributions/1660/attachments/1000/1391/talk-nobackup.pdf
[37] G. Colangelo et al., Phys. Lett. B **833**, 137313 (2022), 2205.12963
[38] S. Gottlieb (2022), https://www.benasque.org/2022lattice_workshop/talks_contr/158_Gottlieb_gm2_LatticeNET.pdf
[39] C. Aubin, T. Blum, M. Golterman, S. Peris, Phys. Rev. D **106**, 054503 (2022), 2204.12256
[40] G. Wang, T. Draper, K.F. Liu, Y.B. Yang (χQCD) (2022), 2204.01280
[41] C. Lehner, A.S. Meyer, Phys. Rev. D **101**, 074515 (2020), 2003.04177
[42] E.H. Chao et al., Eur. Phys. J. C **81**, 651 (2021), 2104.02632
[43] E.H. Chao, R.J. Hudspith, A. Gérardin, J.R. Green, H.B. Meyer, Eur. Phys. J. C **82**, 664 (2022), 2204.08844
[44] J. Leutgeb, A. Rebhan, Phys. Rev. D **101**, 114015 (2020), 1912.01596
[45] L. Cappiello, O. Catà, G. D'Ambrosio, D. Greynat, A. Iyer, Phys. Rev. D **102**, 016009 (2020), 1912.02779
[46] M. Hoferichter, P. Stoffer, JHEP **05**, 159 (2020), 2004.06127
[47] J. Lüdtke, M. Procura, Eur. Phys. J. C **80**, 1108 (2020), 2006.00007
[48] J. Bijnens, N. Hermansson-Truedsson, L. Laub, A. Rodríguez-Sánchez, JHEP **10**, 203 (2020), 2008.13487
[49] J. Bijnens, N. Hermansson-Truedsson, L. Laub, A. Rodríguez-Sánchez, JHEP **04**, 240 (2021), 2101.09169
[50] M. Zanke, M. Hoferichter, B. Kubis, JHEP **07**, 106 (2021), 2103.09829
[51] I. Danilkin, M. Hoferichter, P. Stoffer, Phys. Lett. B **820**, 136502 (2021), 2105.01666
[52] G. Colangelo, F. Hagelstein, M. Hoferichter, L. Laub, P. Stoffer, Eur. Phys. J. C **81**, 702 (2021), 2106.13222
[53] S. Holz, C. Hanhart, M. Hoferichter, B. Kubis, Eur. Phys. J. C **82**, 434 (2022), 2202.05846
[54] J. Grange et al. (Muon $g-2$) (2015), 1501.06858
[55] M. Passera, W.J. Marciano, A. Sirlin, Phys. Rev. D **78**, 013009 (2008), 0804.1142
[56] A. Crivellin, M. Hoferichter, C.A. Manzari, M. Montull, Phys. Rev. Lett. **125**, 091801 (2020), 2003.04886
[57] A. Keshavarzi, W.J. Marciano, M. Passera, A. Sirlin, Phys. Rev. D **102**, 033002 (2020), 2006.12666
[58] B. Malaescu, M. Schott, Eur. Phys. J. C **81**, 46 (2021), 2008.08107
[59] G. Colangelo, M. Hoferichter, P. Stoffer, Phys. Lett. B **814**, 136073 (2021), 2010.07943
[60] M. Cè et al., JHEP **08**, 220 (2022), 2203.08676
[61] D. Boito, M. Golterman, K. Maltman, S. Peris (2022), 2210.13677
[62] C. Bouchiat, L. Michel, J. Phys. Radium **22**, 121 (1961)
[63] S.J. Brodsky, E. de Rafael, Phys. Rev. **168**, 1620 (1968)
[64] J. Calmet, S. Narison, M. Perrottet, E. de Rafael, Phys. Lett. B **61**, 283 (1976)
[65] B. Krause, Phys. Lett. B **390**, 392 (1997), hep-ph/9607259
[66] K. Hagiwara, A.D. Martin, D. Nomura, T. Teubner, Phys. Rev. D **69**, 093003 (2004), hep-ph/0312250
[67] F. Jegerlehner, *The Anomalous Magnetic Moment of the Muon*, Vol. 274 (Springer, Cham, 2017)
[68] M. Hoferichter, T. Teubner, Phys. Rev. Lett. **128**, 112002 (2022), 2112.06929
[69] M. Hoferichter et al., PoS **LATTICE2022**, 316 (2022), 2210.11904
[70] C.L. James, R. Lewis, K. Maltman, Phys. Rev. D **105**, 053010 (2022), 2109.13729
[71] G. Colangelo, M. Hoferichter, B. Kubis, P. Stoffer, JHEP **10**, 032 (2022), 2208.08993
[72] G. Colangelo, M. Hoferichter, B. Kubis, M. Niehus, J. Ruiz de Elvira, Phys. Lett. B **825**, 136852 (2022), 2110.05493
[73] D. Stamen, D. Hariharan, M. Hoferichter, B. Kubis, P. Stoffer, Eur. Phys. J. C **82**, 432 (2022), 2202.11106
[74] B. Aubert et al. (BaBar), Phys. Rev. Lett. **103**, 231801 (2009), 0908.3589
[75] J.P. Lees et al. (BaBar), Phys. Rev. D **86**, 032013 (2012), 1205.2228
[76] F. Ambrosino et al. (KLOE), Phys. Lett. B **670**, 285 (2009), 0809.3950
[77] F. Ambrosino et al. (KLOE), Phys. Lett. B **700**, 102 (2011), 1006.5313
[78] D. Babusci et al. (KLOE), Phys. Lett. B **720**, 336 (2013), 1212.4524
[79] A. Anastasi et al. (KLOE-2), JHEP **03**, 173 (2018), 1711.03085
[80] S. Actis et al. (Working Group on Radiative Corrections, Monte Carlo Generators for Low Energies), Eur. Phys. J. C **66**, 585 (2010), 0912.0749
[81] A. Hoefer, J. Gluza, F. Jegerlehner, Eur. Phys. J. C **24**, 51 (2002), hep-ph/0107154
[82] H. Czyż, A. Grzelińska, J.H. Kühn, G. Rodrigo, Eur. Phys. J. C **39**, 411 (2005), hep-ph/0404078
[83] J. Gluza, A. Hoefer, S. Jadach, F. Jegerlehner, Eur. Phys. J. C **28**, 261 (2003), hep-ph/0212386

[84] Y.M. Bystritskiy, E.A. Kuraev, G.V. Fedotovich, F.V. Ignatov, Phys. Rev. D **72**, 114019 (2005), hep-ph/0505236

[85] B. Moussallam, Eur. Phys. J. C **73**, 2539 (2013), 1305.3143

[86] G. Rodrigo, H. Czyż, J.H. Kühn, M. Szopa, Eur. Phys. J. C **24**, 71 (2002), hep-ph/0112184

[87] F. Campanario et al., JHEP **02**, 114 (2014), 1312.3610

[88] F. Campanario et al., Phys. Rev. D **100**, 076004 (2019), 1903.10197

[89] F. Ignatov, R.N. Lee, Phys. Lett. B **833**, 137283 (2022), 2204.12235

[90] G. Colangelo, M. Hoferichter, J. Monnard, J. Ruiz de Elvira, JHEP **08**, 295 (2022), 2207.03495

[91] G. Abbiendi et al. (2022), 2201.12102

[92] S. Binner, J.H. Kühn, K. Melnikov, Phys. Lett. B **459**, 279 (1999), hep-ph/9902399

[93] H. Czyż, A. Grzelińska, J.H. Kühn, Phys. Lett. B **611**, 116 (2005), hep-ph/0412239

[94] J.P. Lees et al. (BaBar), Phys. Rev. D **92**, 072015 (2015), 1508.04008

[95] J. Monnard, Ph.D. thesis, Bern University (2020), https://boristheses.unibe.ch/2825/

[96] R. Alemany, M. Davier, A. Hoecker, Eur. Phys. J. C **2**, 123 (1998), hep-ph/9703220

[97] S. Schael et al. (ALEPH), Phys. Rept. **421**, 191 (2005), hep-ex/0506072

[98] M. Fujikawa et al. (Belle), Phys. Rev. D **78**, 072006 (2008), 0805.3773

[99] V. Cirigliano, G. Ecker, H. Neufeld, Phys. Lett. B **513**, 361 (2001), hep-ph/0104267

[100] V. Cirigliano, G. Ecker, H. Neufeld, JHEP **08**, 002 (2002), hep-ph/0207310

[101] M. Davier et al., Eur. Phys. J. C **66**, 127 (2010), 0906.5443

[102] F. Jegerlehner, R. Szafron, Eur. Phys. J. C **71**, 1632 (2011), 1101.2872

[103] W. Altmannshofer et al. (Belle-II), PTEP **2019**, 123C01 (2019), [Erratum: PTEP **2020**, 029201 (2020)], 1808.10567

[104] M. Bruno (2022), https://indico.ph.ed.ac.uk/event/112/contributions/1684/attachments/1029/1430/edi22-bruno.pdf

[105] G. Colangelo et al. (2022), 2203.15810

[106] M.N. Achasov et al. (SND), JHEP **01**, 113 (2021), 2004.00263

[107] M. Ablikim et al. (BESIII) (2019), 1912.11208

[108] J.P. Lees et al. (BaBar), Phys. Rev. D **104**, 112003 (2021), 2110.00520

[109] M. Ablikim et al. (BESIII), Phys. Rev. Lett. **128**, 062004 (2022), 2112.11728

[110] A.E. Ryzhenenkov et al., J. Phys. Conf. Ser. **1526**, 012009 (2020)

[111] M. Ablikim et al. (BESIII), Chin. Phys. C **44**, 040001 (2020), 1912.05983

[112] C.M. Carloni Calame, M. Passera, L. Trentadue, G. Venanzoni, Phys. Lett. B **746**, 325 (2015), 1504.02228

[113] G. Abbiendi et al., Eur. Phys. J. C **77**, 139 (2017), 1609.08987

[114] G. Abbiendi et al. (The MUonE Collaboration), Tech. Rep. CERN-SPSC-2019-026. SPSC-I-252, CERN, Geneva (2019), https://cds.cern.ch/record/2677471

[115] P. Banerjee et al., Eur. Phys. J. C **80**, 591 (2020), 2004.13663

[116] L. Di Luzio, A. Masiero, P. Paradisi, M. Passera, Phys. Lett. B **829**, 137037 (2022), 2112.08312

[117] L. Darmé, G. Grilli di Cortona, E. Nardi, JHEP **06**, 122 (2022), 2112.09139

[118] A. Crivellin, M. Hoferichter (2022), 2211.12516

Leading hadronic contribution to the muon g-2 from lattice QCD

B. C. Toth[1,*] for the Budapest–Marseille–Wuppertal collaboration

[1] *University of Wuppertal*

Abstract. We compute the leading order hadronic vacuum polarization contribution to the anomalous magnetic moment of the muon. The calculations are performed using four flavors of stout smeared staggered quarks, with quark masses at their physical values. The continuum limit is taken using six different lattice spacings ranging from 0.132 fm down to 0.064 fm. All strong isospin breaking and electromagnetic effects are accounted for to leading order. A controlled infinite volume limit is taken thanks to dedicated simulations performed in box sizes up to 11 fm. Putting all these ingredients together, we find $[(g_\mu - 2)/2]_{\text{LO-HVP}} = 707.5[5.5] \times 10^{-10}$, which has a total uncertainty of 0.8%. Compared to determinations based on the $e^+e^- \to$ hadrons cross section, our result significantly reduces the tension between the standard model prediction for the muon $g-2$ and its experimental value.

1 Introduction

The muon is a 207 times heavier sibling of the electron, it has the same electric charge and spin. Similarly to the electron, the muon has a magnetic moment, which is proportional to its spin and charge, and inversely proportional to twice its mass. The Dirac equation of relativistic quantum mechanics predicts that the constant of proportionality, g_μ, should be 2. However, in quantum field theory this prediction receives small corrections due to vacuum fluctuations. These corrections are quantified by the anomalous magnetic moment $a_\mu = (g_\mu - 2)/2$. It was measured to a precision of 0.54 ppm at the Brookhaven National Laboratory in the early 2000s [1], and was confirmed recently by Fermilab [2].

At this level of precision, all of the interactions of the standard model are relevant. Although the leading contributions are described by quantum electrodynamics (QED), the one that dominates the theory error is induced by the strong interaction and requires solving the highly non-linear equations of quantum chromodynamics (QCD) at low energies. This contribution is determined by the leading-order, hadronic vacuum polarization (LO-HVP), which describes how the propagation of a virtual photon is modified by the presence of quark and gluon fluctuations in the vacuum. Here we compute this LO-HVP contribution $a_\mu^{\text{LO-HVP}}$, using ab initio simulations in QCD and QED. In the present work, we include both QED and QCD in a lattice formulation, including four non-degenerate quark flavors (up, down, strange and charm). We also consider the tiny contributions of the bottom and top quarks.

We compute $a_\mu^{\text{LO-HVP}}$ in the time-momentum representation [3], therefore we need the following two-point function in Euclidean time t:

$$G(t) = \frac{1}{3e^2} \sum_{\mu=1,2,3} \int d^3x \langle J_\mu(\vec{x}, t) J_\mu(0) \rangle \,, \tag{1}$$

where J_μ is the electromagnetic current of quarks with $J_\mu/e = \frac{2}{3}\bar{u}\gamma_\mu u - \frac{1}{3}\bar{d}\gamma_\mu d - \frac{1}{3}\bar{s}\gamma_\mu s + \frac{2}{3}\bar{c}\gamma_\mu c$. u, d, s and c are the up, down, strange and charm quark fields and the angle brackets stand for the QCD+QED expectation value to order e^2. It is useful to decompose $G(t)$ into light, strange, charm and disconnected components, since they have very different statistical and systematic uncertainties. Integrating the one-photon-irreducible (1γI) part of the two-point function (1) yields the LO-HVP contribution to the magnetic moment of the muon [3–6]:

$$a_\mu^{\text{LO-HVP}} = \alpha^2 \int_0^\infty dt\, K(t)\, G_{1\gamma\text{I}}(t) \,, \tag{2}$$

with the weight function,

$$K(t) = \int_0^\infty \frac{dQ^2}{m_\mu^2}\, \omega\!\left(\frac{Q^2}{m_\mu^2}\right) \left[t^2 - \frac{4}{Q^2}\sin^2\!\left(\frac{Qt}{2}\right)\right] , \tag{3}$$

and where $\omega(r) = [r + 2 - \sqrt{r(r+4)}]^2/\sqrt{r(r+4)}$, α is the fine structure constant in the Thomson limit and m_μ is the muon mass. We are computing only the LO-HVP contribution, therefore we drop the superscript for brevity and multiply the result by 10^{10}.

The subpercent precision, that we are aiming for, represents a huge challenge for lattice QCD. In order to reach that goal, we have to address the following critical issues: scale determination; QED and strong-isospin breaking; noise reduction; infinite-volume and continuum extrapolations. We briefly discuss these one by one. For a more detailed exposition we refer the reader to the Supplementary Information in [7].

*e-mail: btoth@uni-wuppertal.de

2 Scale determination

The quantity a_μ depends on the muon mass. When computing (2) on the lattice, m_μ has to be converted into lattice units, am_μ, where a is the lattice spacing. A relative error of the lattice spacing propagates into about a twice as large a relative error on a_μ, so that a has to be determined with a few permil precision. We use the mass of the Ω^- baryon, $M_{\Omega^-} = 1672.45(29)$ MeV [8], to set the lattice spacing.

To extract the mass of the positive-parity, ground-state Ω baryon, we are using the operator [9]

$$\Omega_{\rm VI}(t) = \sum_{x_4 \text{ even}} \epsilon_{abc} \left[S_1\chi_a S_{12}\chi_b S_{13}\chi_c - S_2\chi_a S_{21}\chi_b S_{23}\chi_c + S_3\chi_a S_{31}\chi_b S_{32}\chi_c \right](x). \quad (4)$$

Here, $\chi_a(x)$ is the strange-quark field with color index a, and the operator S_μ performs a symmetric, gauge-covariant shift in direction μ, while $S_{\mu\nu} \equiv S_\mu S_\nu$.

On the Ω propagators we perform a four-state fit using the fit function h, with two positive and two negative parity states:

$$h(t, A, M) = A_0 h_+(M_0, t) + A_1 h_-(M_1, t) + A_2 h_+(M_2, t) + A_3 h_-(M_3, t) \quad (5)$$

with $h_+(M,t) = e^{-Mt} + (-1)^{t-1} e^{-M(T-t)}$ and $h_-(M,t) = -h_+(M, T-t)$ describing the time dependence of the positive and negative parity states. Here M_0 and A_0 are the mass and amplitude of the ground state. In order to stabilize the fit, a prior term was introduced, containing priors on the masses except for the ground state. The prior for the negative parity ground state, $M_1 = 2012$ MeV, is motivated by the recent observation from the Belle collaboration [10]. The excited states, $M_2 = 2250$ MeV and $M_3 = 2400$ MeV, have not been discovered in experiments so far, so their priors follow from the quark model [11].

In addition to the above four-state fit to the Ω propagator we also used a mass extraction procedure proposed in [12], which is based on the Generalized Eigenvalue Problem (GEVP). The method has the advantage of not using priors. For each time slice t we construct a 4×4 matrix $\mathcal{H}_{i,j}(t) = H_{t+(i-1)+(j-1)}$ $(i, j = 1, \ldots, 4)$, from the hadron propagator H_t. Then for a given t_a and t_b let $\lambda(t_a, t_b)$ be an eigenvalue and $v(t_a, t_b)$ an eigenvector solution to this 4×4 generalized eigenvalue problem:

$$\mathcal{H}(t_a) v(t_a, t_b) = \lambda(t_a, t_b) \mathcal{H}(t_b) v(t_a, t_b). \quad (6)$$

Here we select the smallest eigenvalue λ and use the corresponding eigenvector v to project out the ground state:

$$v^+(t_a, t_b) \mathcal{H}(t) v(t_a, t_b), \quad (7)$$

which then can be fitted to a simple $\exp(-Mt)$ type function.

The mass extracted using the GEVP gives a third M_Ω value for each ensemble, beside the results with the four-state fit procedure with two fit ranges. We will use the deviation between these three values as a systematic error in the Ω mass determination.

3 Isospin-breaking effects

Our staggered path integral includes four flavors of quarks, $f = \{u, d, s, c\}$, gluon fields U and photon fields A and is given by:

$$Z = \int [dU] \exp(-S_g[U]) \int [dA] \exp(-S_\gamma[A]) \times \prod_f \det M^{1/4}[V_U \exp(ieq_f A), m_f]. \quad (8)$$

The photon integral measure $[dA]$ and action S_γ are defined in the QED$_{\rm L}$ scheme [13]. The one-hop staggered matrix in a background field W_μ can be written as

$$M[W, m] = D[W] + m = \sum_\mu D_\mu[W_\mu] + m, \quad (9)$$

where D_μ is the covariant differentiation in the μ direction involving W and its adjoint $W^\dagger$ together with the obligatory staggered phases. In the path integral the fermions are coupled to a gauge field that is a product of the exponentiated photon field and of the smeared gluon gauge field V_U. $q_f \in \{+\frac{2}{3}, -\frac{1}{3}, -\frac{1}{3}, +\frac{2}{3}\}$ stand for the quark electric charges in units of the positron charge e, m_f for the quark masses and $\alpha = e^2/(4\pi)$. We use the notation $\delta m \equiv m_d - m_u$ for the difference in the up and down quark masses and $m_l \equiv \frac{1}{2}(m_u + m_d)$ for their average. To simplify later formulas we also introduce the notations

$$M_f \equiv M[V_U e^{ieq_f A}, m_f], \quad (10)$$

$$\text{dets}[U, A; \{m_f\}, \{q_f\}, e] \equiv \prod_f \det M_f^{1/4}, \quad (11)$$

where the latter is the product of all fermion determinants.

In this work isospin-breaking is implemented by taking derivatives with respect to the isospin-breaking parameters and by measuring the so obtained derivative operators on isospin-symmetric configurations [14]. We introduce a set of notations for isospin-symmetric observables and their isospin-breaking derivatives. Consider the observable $X(e, \delta m)$, which is a function of e and δm. Then we define

$$X_0 \equiv X(0,0), \qquad X'_m \equiv m_l \frac{\partial X}{\partial \delta m}(0,0),$$
$$X'_1 \equiv \frac{\partial X}{\partial e}(0,0), \qquad X''_2 \equiv \frac{1}{2}\frac{\partial^2 X}{\partial e^2}(0,0). \quad (12)$$

We take into account only leading-order isospin-breaking in this work, so no higher derivatives are needed.

In the case of the fermion determinant, the isospin-symmetric value is denoted by dets$_0$. Since dets is symmetric under the exchange $u \leftrightarrow d$, the strong-isospin-breaking of dets is zero at leading order: $\text{dets}'_m = 0$. The electromagnetic derivatives are

$$\frac{\text{dets}'_1}{\text{dets}_0} = \sum_f \frac{q_f}{4} \text{Tr}\left(M_f^{-1} D[iAV_U]\right), \quad (13)$$

$$\frac{\text{dets}''_2}{\text{dets}_0} = \frac{1}{2}\left[\left(\frac{\text{dets}'_1}{\text{dets}_0}\right)^2 - \sum_f \frac{q_f^2}{4} \text{Tr}\left(M_f^{-1} D[A^2 V_U]\right) - \sum_f \frac{q_f^2}{4} \text{Tr}\left(M_f^{-1} D[iAV_U] M_f^{-1} D[iAV_U]\right)\right],$$

where Tr is trace over color and spacetime indices and the argument of the D operator is a 3×3 complex matrix valued field, e.g. $A^2 V_U$ has components $A^2_{\mu,x}[V_U]_{\mu,x}$.

We also make a distinction between the electric charge in the fermion determinant and in the operator that we measure. We call the former sea electric charge and denote it by e_s, the latter is the valence electric charge and is denoted by e_v. For an observable X that depends on both the valence and sea charges, $X(e_v, e_s)$, the second order electric charge derivatives are defined as $X''_{20} \equiv \frac{1}{2}\frac{\partial^2 X}{\partial e_v^2}(0,0)$, $X''_{11} \equiv \frac{\partial^2 X}{\partial e_v \partial e_s}(0,0)$, $X''_{02} \equiv \frac{1}{2}\frac{\partial^2 X}{\partial e_s^2}(0,0)$. For functions that depend on either e_v or e_s, but not on both, we use the single digit notations of Equation (3).

The expectation value of an operator O is calculated by inserting $O[U,A]$ into the integrand of the path integral of Equation (8) and normalizing the integral by Z. Here we consider operators whose photon field dependence arises entirely from the photon-quark interaction, i.e. $O = O[U, e_v A]$. The expectation value of this operator depends on δm, e_v and e_s, and the expansion in isospin breaking corrections can be written as:

$$\langle O \rangle = [\langle O \rangle]_0 + e_v^2 \langle O \rangle''_{20} + e_v e_s \langle O \rangle''_{11} + e_s^2 \langle O \rangle''_{02} + \frac{\delta m}{m_l} \langle O \rangle'_m . \tag{14}$$

Here, the individual terms can be expressed as expectation values obtained with the isospin-symmetric path integral, which we denote by $\langle \ldots \rangle_0$. The concrete expressions are:

$$\begin{aligned}
\text{iso:} \quad & [\langle O \rangle]_0 = \langle O_0 \rangle_0 \\
\text{val-val:} \quad & \langle O \rangle''_{20} = \langle O''_2 \rangle_0 \\
\text{sea-val:} \quad & \langle O \rangle''_{11} = \left\langle O'_1 \frac{\text{dets}'_1}{\text{dets}_0} \right\rangle_0 \\
\text{sea-sea:} \quad & \langle O \rangle''_{02} = \left\langle O_0 \frac{\text{dets}''_2}{\text{dets}_0} \right\rangle_0 - \langle O_0 \rangle_0 \left\langle \frac{\text{dets}''_2}{\text{dets}_0} \right\rangle_0 \\
\text{sib:} \quad & \langle O \rangle'_m = \langle O'_m \rangle_0
\end{aligned} \tag{15}$$

In the derivation of these expressions we use $\left\langle \frac{\text{dets}'_1}{\text{dets}_0} \right\rangle_0 = 0$.

Note that Equation (14) is an expansion in bare parameters and not what we consider a decomposition into isospin-symmetric and isospin breaking parts. The latter involves derivatives with respect to renormalized observables and our prescription for that is given in Section 3.1. There is no need to introduce a renormalized electromagnetic coupling though: its running is an $O(e^4)$ effect, i.e. beyond the leading order isospin approximation that we consider here.

3.1 Isospin-breaking decomposition

For various purposes it is useful to decompose the observables into isospin-symmetric and isospin-breaking parts. This requires a matching of the isospin symmetric and full theories, in which we specify a set of observables that must be equal in both theories. Of course, different sets will lead to different decompositions, which is commonly referred to as scheme dependence. Only the sum of the components, i.e. the result in the full theory, is scheme independent.

A possible choice for the observables are the Wilson-flow–based w_0 scale and the masses of mesons built from an up/down/strange and an anti-up/down/strange quark, $M_{uu}/M_{dd}/M_{ss}$. These mesons are defined by taking into account only the quark-connected contributions in their two-point functions [15]. Their masses are practical substitutes for the quark masses. Also, they are neutral and have no magnetic moment, so they are a reasonable choice for an isospin decomposition. These masses cannot be measured in experiments, but have a well defined continuum limit and thus a physical value can be associated to them. In particular, we use the combinations $M^2_{\pi_\chi} \equiv \frac{1}{2}(M^2_{uu} + M^2_{dd})$ and $\Delta M^2 \equiv M^2_{dd} - M^2_{uu}$. For the determination of the physical values of w_0, M_{ss} and ΔM^2, see the Supplementary Information of [7].

For the decomposition we start with the QCD+QED theory and parameterize our observable $\langle O \rangle$ with the quantities defined above:

$$\langle O \rangle (M_{\pi_\chi} w_0, M_{ss} w_0, \tfrac{L}{w_0}, \Delta M w_0, e) . \tag{16}$$

Here, the continuum limit is assumed. We can isolate the electromagnetic part by switching off the electromagnetic coupling, while keeping the other parameters fixed. The strong-isospin-breaking part is given by the response to the ΔM parameter, and the isospin-symmetric part is just the remainder:

$$\langle O \rangle_{\text{qed}} \equiv e^2 \cdot \left. \frac{\partial \langle O \rangle}{\partial e^2} \right|_{M_{\pi_\chi} w_0, M_{ss} w_0, \frac{L}{w_0}, \Delta M w_0, e=0} \tag{17}$$

$$\langle O \rangle_{\text{sib}} \equiv (\Delta M w_0)^2 \cdot \left. \frac{\partial \langle O \rangle}{\partial (\Delta M w_0)^2} \right|_{M_{\pi_\chi} w_0, M_{ss} w_0, \frac{L}{w_0}, \Delta M w_0 = 0, e=0} \tag{18}$$

$$\langle O \rangle_{\text{iso}} \equiv \langle O \rangle (M_{\pi_\chi} w_0, M_{ss} w_0, \tfrac{L}{w_0}, 0, 0). \tag{19}$$

One can also define the decomposition at a finite lattice spacing, for which w_0 in lattice units can be additionally fixed. In doing so the isospin symmetric part $\langle O \rangle_{\text{iso}}$ has to be distinguished from the value of the observable at the bare isospin-symmetric point $[\langle O \rangle]_0$.

4 Noise reduction techniques

In this section we describe the measurement techniques that were implemented for the isospin symmetric component of the connected light quark contribution, C_0^{light}. They play a key role in reducing the final statistical error in a_μ. For further details we refer to Ref. [7].

4.1 Low Mode Averaging

The technique utilizes the lowest eigenmodes of the fermion matrix; for an early work with low eigenmodes, see [16]. The way in which we use these modes here is essentially the same as in [17], where it is called Low Mode Substitution. In the space orthogonal to these modes, the computational effort is reduced considerably by applying imprecise (aka. sloppy) matrix inversions. This is called the Truncated Solver Method [18] or All Mode Averaging [19].

We consider the connected current propagator for timelike separation, and perform an averaging over the source positions, together with a zero spatial-momentum projection at the sink:

$$C(t,\bar{t}) = -\frac{1}{12L^3}\sum_{\mu=1,2,3}\mathrm{ReTr}\left[\mathcal{D}_{\mu,t}M^{-1}\mathcal{D}_{\mu,\bar{t}}M^{-1}\right], \quad (20)$$

where $\mathcal{D}_{\mu,t} = \sum_{\vec{x}} D_\mu[iP_x U]$ is an operator that performs a symmetric, gauge-covariant shift on a vector v_x:

$$[\mathcal{D}_{\mu,t}v]_x = i\eta_{\mu,x}\left(U_{\mu,x}v_{x+\mu} + U^\dagger_{\mu,x-\mu}v_{x-\mu}\right)\cdot\delta_{x_4,t}\,, \quad (21)$$

where $\eta_{\mu,x}$ are the staggered phases. We use the simplifying notation $\mathcal{D} = \mathcal{D}_{\mu,t}$ and $\overline{\mathcal{D}} = \mathcal{D}_{\mu,\bar{t}}$ in the following. In Equation (20), we apply the real part to reduce noise, because the imaginary part vanishes anyway after averaging over gauge configurations.

Using the lowest eigenmodes of M we split the quark propagator into an eigenvector part and into its orthogonal complement, $M^{-1} = M_e^{-1} + M_r^{-1}$, with

$$M_e^{-1} = \sum_i \frac{1}{\lambda_i}v_i v_i^\dagger, \quad M_r^{-1} = M^{-1}\left(1 - \sum_i v_i v_i^\dagger\right), \quad (22)$$

where v_i/λ_i is the i-th eigenvector/eigenvalue of the operator M. Correspondingly, C splits into eigen-eigen, rest-eigen and rest-rest contributions:

$$C = C_{ee} + C_{re} + C_{rr} = -\frac{1}{4L^3}\sum_{\substack{p=e,r\\q=e,r}}\mathrm{ReTr}\left[\mathcal{D}M_p^{-1}\overline{\mathcal{D}}M_q^{-1}\right], \quad (23)$$

where an average over μ is assumed but not shown explicitely. The benefit of this decomposition is that the trace in the **eigen-eigen part** can be calculated exactly, and is thus equivalent to calculating the propagator with all possible sources in position space. This is the main ingredient for the noise reduction. Though no extra inversions are needed in this part, it has to be optimized carefully, since there is a double sum over the eigenmodes, where each term is a scalar product $v_i^\dagger \mathcal{D} v_j$. In the **rest-eigen part** we have terms $v_i^\dagger \mathcal{D}M_r^{-1}\overline{\mathcal{D}}v_i$ and also terms where $\mathcal{D}$ and $\overline{\mathcal{D}}$ are exchanged. Therefore, this part is only a single sum over the eigenmodes, and each term involves one matrix inversion. Note that these inversions are preconditioned by the eigenvectors, so they need many fewer iterations than standard inversions. Additionally, we speed up the inversions by running them with a reduced precision, and for some randomly selected eigenvectors we correct for the small bias by adding the difference between a high precision solver and the reduced precision one [18, 19]. Finally, the **rest-rest part** is evaluated using random source vectors ξ: we calculate $\xi^\dagger\mathcal{D}M_r^{-1}\overline{\mathcal{D}}M_r^{-1}\xi$, which requires two inversions per random source. The reduced precision inverter technique is used here too. The improvement achieved by these noise reduction techniques is shown in Figure 1.

4.2 Upper and lower bounds on $\langle JJ\rangle$

In the case of the light and disconnected contributions to the current propagator, the signal deteriorates quickly as distance is increased. To calculate the HVP, a sum over time of the propagator has to be performed. As was suggested in [21, 22], we introduce a cut in time t_c, beyond which the propagator is replaced by upper and lower bounds, thereby reducing the statistical noise. Our estimate is given by the average of the bounds at a t_c where the two bounds meet. The bounds are derived from the fact that the current propagator is a sum of exponentials with positive coefficients.

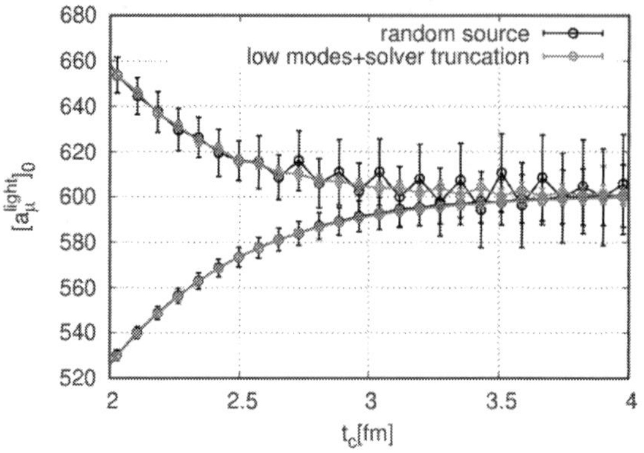

Figure 1. Comparison of a conventional random source based technique, as we applied it in our earlier work [20], and the low-mode improvement technique of this work on a β = 3.9200 `4stout` ensemble for the case of $[a_\mu^{\mathrm{light}}]_0$'s upper and lower bounds (see Section 4.2).

For the light connected propagator at the isospin-symmetric point the bounds express the positivity (lower bound) and that the propagator should decay faster than the exponential of two pions (upper bound). They are given as

$$0 \le G^{\mathrm{light}}(t) \le G^{\mathrm{light}}(t_c)\frac{\varphi(t)}{\varphi(t_c)}, \quad (24)$$

where $\varphi(t) = \exp(-E_{2\pi}t)$. For $E_{2\pi}$ we use the energy of two non-interacting pions with the smallest non-zero lattice momentum $2\pi/L$. The larger the t_c the better the upper bound, but it comes with more statistical noise.

The exponential decay above assumes an infinite time extent, $T = \infty$. We incorporate the effects of a finite-T using next-to-leading-order chiral perturbation theory. There the exponential decay with the two-pion energy gets replaced by the following cosh-type form:

$$\exp(-E_{2\pi}t) \longrightarrow \frac{\cosh[E_{2\pi}(t-T/2)]+1}{\cosh(E_{2\pi}T/2)-1}\,. \quad (25)$$

5 Finite-size effects

We compute finite-size effects on a_μ in a systematic way, which includes dedicated lattice simulations, chiral perturbation theory and phenomenological models. The concrete goal is to provide a single number that is to be added to the continuum-extrapolated lattice result obtained in a reference box, which is defined by a spatial extent of $L_{\rm ref} = 6.272$ fm and a temporal extent of $T_{\rm ref} = \frac{3}{2}L_{\rm ref}$.

We perform dedicated lattice simulations with two different lattice geometries: one is a 56×84 lattice with the reference box size and the other is a large 96×96 lattice with box size $L = L_{\rm big} = 10.752$ fm and $T = T_{\rm big} = L_{\rm big}$. Since taste violations distort the finite-size effects, we designed a new action with highly-suppressed taste breaking, which we call 4HEX (see [7]). Our strategy is then to compute the finite-size correction as the following sum:

$$\begin{aligned} a_\mu(\infty,\infty) - a_\mu(L_{\rm ref}, T_{\rm ref}) = \\ = [a_\mu(L_{\rm big}, T_{\rm big}) - a_\mu(L_{\rm ref}, T_{\rm ref})]_{\rm 4HEX} \\ + [a_\mu(\infty,\infty) - a_\mu(L_{\rm big}, T_{\rm big})]_{\rm XPT}. \end{aligned} \quad (26)$$

The first difference on the right hand side is taken from the dedicated 4HEX simulations. The second difference is expected to be much smaller than the first and is taken from a non-lattice approach: two-loop chiral perturbation theory.

We consider four non-lattice approaches to compute both differences on the right hand side of Equation (26). In the case of the first difference, the results obtained are compared to our 4HEX simulations. The first approach is chiral perturbation theory (XPT) to next-to-leading and next-to-next-to-leading orders (NLO and NNLO), the second is the Meyer-Lellouch-Luscher-Gounaris-Sakurai model (MLLGS), the third approach is that of Hansen and Patella (HP) [23] and the fourth is the rho-pion-gamma model of [24], which we abbreviate as RHO here.

We compute the first difference in Equation (26) using dedicated simulations with the 4HEX action. We use the harmonic-mean-square (HMS) to set the physical point:

$$M_{\pi,\rm HMS}^{-2} \equiv \frac{1}{16}\sum_\alpha M_{\pi,\alpha}^{-2}\,, \quad (27)$$

defined as an average over the masses of the 16 pion tastes, $M_{\pi,\alpha}$. We set $M_{\pi,\rm HMS}$ to the physical value of the pion mass, which requires lowering the Goldstone-pion mass to 110 MeV. This way of fixing the physical point results in much smaller lattice artefacts than the usual setting with the Goldstone-pion, at least for an observable like the finite-size effect. To generate the 4HEX data set, we performed simulations with two different Goldstone pion masses: $M_\pi = 104$ MeV and 121 MeV. To set the physical point as described above, we perform an interpolation from these two pion masses to $M_\pi = 110$ MeV.

We only have one lattice spacing with the 4HEX action, so the finite-size effects cannot be extrapolated to the continuum limit. We estimate the cutoff effect of the result by comparing $a_\mu^{\rm light}$ with the 4HEX action at this single lattice spacing to the continuum extrapolated `4stout` lattice result, both in the $L_{\rm ref}$ volume. The 4HEX result is about 7% larger than the continuum value. Therefore we reduce the measured finite-size effect by 7%, and assign a 7% uncertainty to this correction step. For the difference we get

$$a_\mu(L_{\rm big}, T_{\rm big}) - a_\mu(L_{\rm ref}, T_{\rm ref}) = 18.1(2.0)_{\rm stat}(1.4)_{\rm cont}\,. \quad (28)$$

Table 1. The finite size effects $a_\mu(\text{big}) - a_\mu(\text{ref})$ for the isospin-symmetric component in the various model approaches.

NLO XPT	NNLO XPT	MLLGS	HP	RHO
11.6	15.7	17.8	16.7	15.2

The result includes a $\left(\frac{9}{10}\right)$ charge factor, and the first error is statistical and the second is an estimate of the cutoff effect.

The finite-size effects computed in various non-lattice approaches are collected in Table 1. Except for the NLO result, the different models give finite-size effects of similar size, which agree well with the lattice determination of Equation (28). The good agreement for the finite-size effect in the reference box, between the models and the lattice, gives us confidence that the models can be used to reliably compute the very small, residual, finite-size effect of the large box. As a final value for the large-box, finite-size effect we take the NNLO XPT result including finite-T effects:

$$a_\mu(\infty,\infty) - a_\mu(L_{\rm big}, T_{\rm big}) = 0.6(0.3)_{\rm big}\,, \quad (29)$$

where the uncertainty is an estimate of higher-order effects, given here by the difference of the NNLO and NLO values.

For our final result for the finite-size effect in the reference box, we also include the contribution of isoscalar channel and isospin-breaking effects giving:

$$\begin{aligned} a_\mu(\infty,\infty) - a_\mu(L_{\rm ref}, T_{\rm ref}) = \\ = 18.7(2.0)_{\rm stat}(1.4)_{\rm cont}(0.3)_{\rm big}(0.6)_{I=0}(0.1)_{\rm qed}[2.5]\,. \end{aligned} \quad (30)$$

The first error is the statistical uncertainty of our 4HEX computation, the second is an estimate of the 4HEX cutoff effects, the third is the uncertainty of the residual finite-size effect of the "big" lattice, the fourth is a XPT estimate of the $I = 0$ finite size effect and the fifth is an estimate of the isospin-breaking effects. The last, total error in the square-brackets is the sum of the first five, added in quadrature.

6 Taste improvement

As is well known, some of the most important cutoff effects of staggered fermions are taste violations. At long distances, these violations distort the pion spectrum. Since a_μ is predominantly a long-distance observable, dominated by a two-pion contribution, including the ρ resonance, we expect these effects to be largest in the light-quark terms.

We investigate various physically motivated models for reducing long-distance taste violations in our lattice results. We consider three techniques: next-to-next-to-leading order chiral perturbation theory (NNLO XPT), a Meyer-Lellouch-Lüscher-Gounaris-Sakurai model (MLLGS) and the rho-pion-gamma model (RHO). A detailed exposition of these models can be found in the Supplementary Material of Ref. [7]. We investigate and discuss the suitability of their staggered versions for reducing the taste violations present in our lattice data. We call the resulting corrections taste improvements, because they improve the continuum extrapolation of our lattice data without, in principle, modifying the continuum-limit value. Indeed, these corrections vanish in that limit, as taste-breaking effects should. These improvements are applied on light-quark observables at the isospin-symmetric point, whose taste violations have the largest impact on our final uncertainties.

The models NNLO XPT, MLLGS and RHO describe the long-distance physics associated with finite-volume effects, as measured in our simulations. One can also define corresponding models describing the taste violations, they are denoted NNLO SXPT, SMLLGS and SRHO. We find that they describe the physics associated with taste violations, at least at larger distances. This is illustrated in Figure 2, where cutoff effects in the integrand of a_μ^{light} are plotted as a function of Euclidean time. More specifically, we define the physical observable, obtained by convoluting the integrand of a_μ^{light} with a smooth window function

$$W(t; t_1, t_2) \equiv \Theta(t; t_1, \Delta) - \Theta(t; t_2, \Delta) \tag{31}$$

with

$$\Theta(t; t', \Delta) \equiv \tfrac{1}{2} + \tfrac{1}{2}\tanh[(t - t')/\Delta], \tag{32}$$

having a width of $t_2 - t_1 = 0.5$ fm and starting at a time of t_1. Then we consider the difference in the value of this observable, obtained on a fine and a coarse lattice at a sequence of t_1 separated by 0.1 fm. These are compared to the NLO SXPT, NNLO SXPT, SRHO and SMLLGS predictions for this quantity, evaluated at the exact parameters of the ensembles.

The SMLLGS, the SRHO and the NNLO SXPT taste improvements describe the numerical data very nicely for $t_1 \gtrsim 2.0$ fm, fairly well for $t_1 \gtrsim 1.0$ fm and all the way down to $t_1 \simeq 0.4$ fm in the case of SRHO. All three slightly overestimate the observed cutoff effects, the rho-meson based approach performing best, whereas NNLO displays a large deviation from the lattice results in the $t_1 \leq 0.8$ fm region. The lattice results have a maximum at $t_1 = 1.4$ fm, as does the SRHO improvement, reinforcing our confidence that this model captures the relevant physics.

These findings lead us to apply the following taste corrections to our simulations results for $a_\mu^{\text{light}}(L, T, a)$, obtained on an $L^3 \times T$ lattice with lattice spacing a, before performing continuum extrapolations:

$$a_\mu^{\text{light}}(L, T, a) \to$$
$$\to a_\mu^{\text{light}}(L, T, a) + \tfrac{10}{9}\left[a_{\mu,t\geq t_{\text{sep}}}^{\text{RHO}}(L_{\text{ref}}, T_{\text{ref}}) - a_{\mu,t\geq t_{\text{sep}}}^{\text{SRHO}}(L, T, a)\right], \tag{33}$$

with $t_{\text{sep}} = 0.4, 0.7, 1.0, 1.3$ fm, and where the factor (10/9) is related to the quark charges. Note that by using L_{ref} and T_{ref} in the above Equation, we are applying a very small volume correction to interpolate all of our simulation results to the same reference, four-volume so that they can be extrapolated to the continuum limit together.

The taste-improved data is then extrapolated to the continuum using our standard fit procedure, in the course

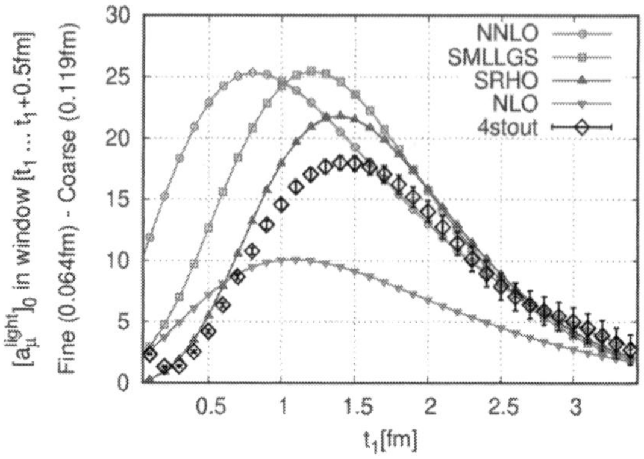

Figure 2. Isospin-symmetric component of a_μ^{light} computed with a sliding window. The window starts at t_1 and ends 0.5 fm later. The plot shows the difference between a fine and a coarse lattice, whose sizes are $L = 6.14$ fm and $L = 6.67$ fm. The black squares with errors are obtained from the simulation. The colored curves are the predictions of NLO and NNLO SXPT, the SRHO and the SMLLGS models. They are computed at the parameters (pion mass, taste violation, volume) of the simulations.

of which isospin-breaking effects are also included. For estimating the systematic error we use a histogram technique [7].

The procedure described above does not yet take into account the systematic uncertainty associated with our choice of SRHO for taste improvement for $t > 1.3$ fm. Since applying no taste improvement in that region is not an option, because of the nonlinearities introduced by two-pion, taste violations, we turn to NNLO SXPT, only as a means to estimate the uncertainty associated with this choice. Thus, we define this systematic uncertainty as ERR = (SRHO − NNLO SXPT) for $t > 1.3$ fm. Then, we perform the same histogram analysis but with SRHO, SRHO-ERR and SRHO+ERR improvements. From this histogram we extract the contribution which comes from the variation in the improvement model from SRHO-ERR to SRHO+ERR. We assign this full spread to the systematic uncertainty associated with the taste-improvement procedure. We add this error in quadrature to the error given by the histogram technique discussed in the previous paragraph.

7 Intermediate window

The work [21] defined a particularly useful observable $a_{\mu,\text{win}}$, in which the current propagator is restricted to a time window $[t_1, t_2]$, using the smooth weight function $W(t; t_1, t_2)$ defined in Equation (31). The advantage of $a_{\mu,\text{win}}$ over a_μ is that, by choosing an appropriate window, the calculation can be made much less challenging on the lattice than for the full a_μ. Here we will be interested in the window between $t_1 = 0.4$ fm and $t_2 = 1.0$ fm, i.e. in an intermediate time range. By this choice we eliminate both the short-distance region, where large cutoff effects are present, and the long-distance region, where the statis-

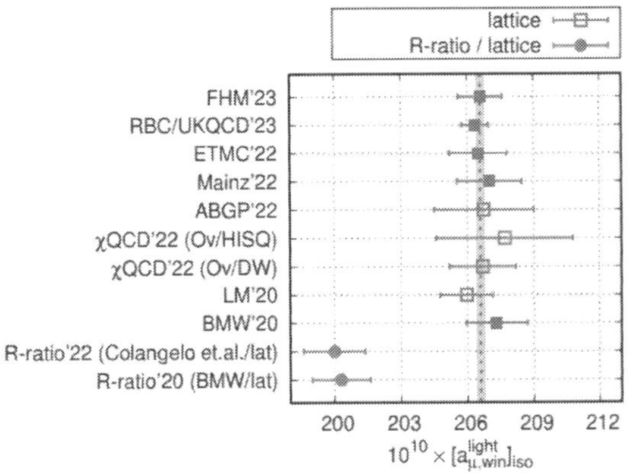

Figure 3. Isospin-symmetric, light, connected component of the window observable, $[a^{\mathrm{light}}_{\mu,\mathrm{win}}]_{\mathrm{iso}}$. The green squares denote the lattice results. From each group only the most recent values are shown: FHM'23 [26], RBC/UKQCD'23 [27], ETMC'22 [28], Mainz'22 [29], ABGP'22 [30], χQCD'22 with overlap valence on HISQ and overlap valence on domain wall configurations [31], LM'20 [32], and our result BMW'20 [7]. The data points with filled squares give a set of independent results: each of these are obtained using a different set of configurations. The green vertical band represents their weighted average. The red dots show the R-ratio based determinations. They are obtained by combining results from the data-driven approach for $a_{\mu,\mathrm{win}}$ and lattice results for the non-light-connected contributions, as described in [7]. The averaged lattice result is 4.8 sigmas away from the value of R-ratio'20 (BMW/lat) [7], and 4.7 sigmas away from the result R-ratio'22 (Colangelo et.al./lat) [25].

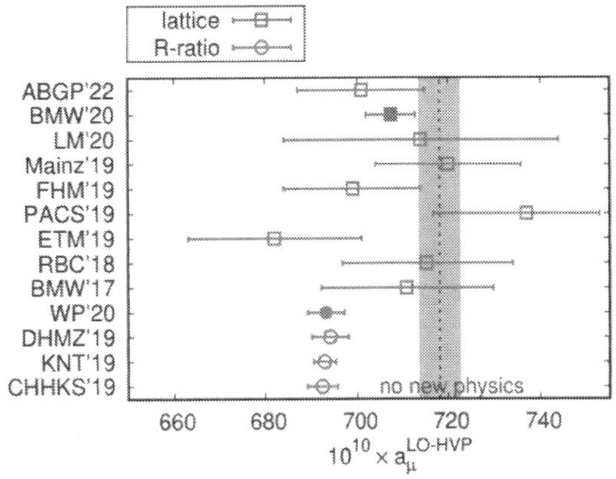

Figure 4. Comparison of recent results for the leading-order, hadronic vacuum polarization contribution to the anomalous magnetic moment of the muon. Green squares are lattice results: this work's result, BMW'20, is represented by a filled symbol. The open squares show other lattice results: ABGP'22 [30], LM'20 [32], Mainz'19 [33], FHM'19 [34], ETM'19 [35], RBC'18 [21] and our earlier work BMW'17 [20]. Red circles were obtained using the R-ratio method: the combined result WP'20 [36] is followed by DHMZ'19 [37], KNT'19 [38] and CHHKS'19 [39, 40]; these results use the same experimental data as input. The blue shaded region is the value that $a_\mu^{\mathrm{LO-HVP}}$ would have to have to explain the experimental measurement of $(g_\mu - 2)$, assuming no new physics.

tical uncertainties and finite-size effects are large. Moreover, in the case of staggered fermions, it has reduced taste-breaking artefacts.

On the other hand, $a_{\mu,\mathrm{win}}$ can be computed in the R-ratio approach [7, 25] as well, which can then be compared to the values obtained on the lattice, providing a sharp comparison of the two methods.

The isospin-symmetric component of the connected, light, window quantity, $[a^{\mathrm{light}}_{\mu,\mathrm{win}}]_{\mathrm{iso}}$, has been computed by several lattice groups. The most recent value from each group is collected in Figure 3. The average of the most recent, independent lattice determinations show a 4.8 σ and a 4.7 σ tension compared to the R-ratio-based determinations of [7] and [25], respectively.

8 Conclusions

Combining all of these ingredients we obtain, as a final result, $a_\mu = 707.5(2.3)(5.0)[5.5]$. The first, statistical error comes mostly from the noisy, large-distance region of the current-current correlator. The second, systematic error is dominated by the continuum extrapolation and the finite-size effect computation. The third, total error is obtained by adding the first two in quadrature. In total we reach a relative accuracy of 0.8%.

Figure 4 compares our result with previous lattice computations and also with results from the R-ratio method, which have recently been reviewed in [36]. As one can see, there is a tension between our result and those obtained by the R-ratio method. For the total, LO-HVP contribution to a_μ, our result is 2.0σ, 2.5σ, 2.4σ and 2.2σ larger than the R-ratio results of $a_\mu = 694.0(4.0)$ [37], $a_\mu = 692.78(2.42)$ [38], $a_\mu = 692.3(3.3)$ [39, 40] and the combined result $a_\mu = 693.1(4.0)$ of [36], respectively. It is worth noting that the R-ratio determinations are based on the same experimental data sets and are therefore strongly correlated, though these data sets were obtained in several different and independent experiments that we have no reason to believe are collectively biased. Clearly, these comparisons need further investigation.

To conclude, when combined with the other standard model contributions (see eg. [36]), our result for the leading-order hadronic contribution to the anomalous magnetic moment of the muon, weakens the longstanding discrepancy between experiment and theory. However, as discussed above and can be seen in Figure 4, our lattice result shows some tension with the R-ratio determinations. Obviously, our findings should be confirmed –or refuted– by other collaborations using other discretizations of QCD.

References

[1] G.W. Bennett et al. (Muon g-2), Phys. Rev. **D73**, 072003 (2006), hep-ex/0602035
[2] B. Abi et al. (Muon g-2), Phys. Rev. Lett. **126**, 141801 (2021), 2104.03281

[3] D. Bernecker, H.B. Meyer, Eur. Phys. J. **A47**, 148 (2011), 1107.4388
[4] B.E. Lautrup, A. Peterman, E. de Rafael, Phys. Rept. **3**, 193 (1972)
[5] E. de Rafael, Phys. Lett. **B322**, 239 (1994), hep-ph/9311316
[6] T. Blum, Phys. Rev. Lett. **91**, 052001 (2003), hep-lat/0212018
[7] S. Borsanyi et al., Nature **593**, 51 (2021), 2002.12347
[8] M. Tanabashi et al. (Particle Data Group), Phys. Rev. **D98**, 030001 (2018)
[9] M.F.L. Golterman, J. Smit, Nucl. Phys. **B255**, 328 (1985)
[10] J. Yelton et al. (Belle), Phys. Rev. Lett. **121**, 052003 (2018), 1805.09384
[11] S. Capstick, N. Isgur, Phys. Rev. **D34**, 2809 (1986), [AIP Conf. Proc.132,267(1985)]
[12] C. Aubin, K. Orginos, AIP Conf. Proc. **1374**, 621 (2011), 1010.0202
[13] M. Hayakawa, S. Uno, Prog. Theor. Phys. **120**, 413 (2008), 0804.2044
[14] G.M. de Divitiis, R. Frezzotti, V. Lubicz, G. Martinelli, R. Petronzio, G.C. Rossi, F. Sanfilippo, S. Simula, N. Tantalo (RM123), Phys. Rev. **D87**, 114505 (2013), 1303.4896
[15] S. Borsanyi et al. (Budapest-Marseille-Wuppertal), Phys. Rev. Lett. **111**, 252001 (2013), 1306.2287
[16] H. Neff, N. Eicker, T. Lippert, J.W. Negele, K. Schilling, Phys. Rev. **D64**, 114509 (2001), hep-lat/0106016
[17] A. Li et al. (xQCD), Phys. Rev. **D82**, 114501 (2010), 1005.5424
[18] G.S. Bali, S. Collins, A. Schafer, Comput. Phys. Commun. **181**, 1570 (2010), 0910.3970
[19] T. Blum, T. Izubuchi, E. Shintani, Phys. Rev. **D88**, 094503 (2013), 1208.4349
[20] S. Borsanyi et al. (Budapest-Marseille-Wuppertal), Phys. Rev. Lett. **121**, 022002 (2018), 1711.04980
[21] T. Blum, P.A. Boyle, V. Gulpers, T. Izubuchi, L. Jin, C. Jung, A. Juttner, C. Lehner, A. Portelli, J.T. Tsang (RBC, UKQCD), Phys. Rev. Lett. **121**, 022003 (2018), 1801.07224
[22] S. Borsanyi, Z. Fodor, T. Kawanai, S. Krieg, L. Lellouch, R. Malak, K. Miura, K.K. Szabo, C. Torrero, B.C. Toth, Phys. Rev. **D96**, 074507 (2017), 1612.02364
[23] M.T. Hansen, A. Patella, Phys. Rev. Lett. **123**, 172001 (2019), 1904.10010
[24] B. Chakraborty, C.T.H. Davies, P.G. de Oliviera, J. Koponen, G.P. Lepage, R.S. Van de Water, Phys. Rev. **D96**, 034516 (2017), 1601.03071
[25] G. Colangelo, A.X. El-Khadra, M. Hoferichter, A. Keshavarzi, C. Lehner, P. Stoffer, T. Teubner, Phys. Lett. B **833**, 137313 (2022), 2205.12963
[26] A. Bazavov et al. (2023), 2301.08274
[27] T. Blum et al. (2023), 2301.08696
[28] C. Alexandrou et al. (2022), 2206.15084
[29] M. Cè et al., Phys. Rev. D **106**, 114502 (2022), 2206.06582
[30] C. Aubin, T. Blum, M. Golterman, S. Peris, Phys. Rev. D **106**, 054503 (2022), 2204.12256
[31] G. Wang, T. Draper, K.F. Liu, Y.B. Yang (chiQCD) (2022), 2204.01280
[32] C. Lehner, A.S. Meyer, Phys. Rev. D **101**, 074515 (2020), 2003.04177
[33] A. Gerardin, M. Ce, G. von Hippel, B. Horz, H.B. Meyer, D. Mohler, K. Ottnad, J. Wilhelm, H. Wittig, Phys. Rev. **D100**, 014510 (2019), 1904.03120
[34] C.T.H. Davies et al. (Fermilab Lattice, LATTICE-HPQCD, MILC), Phys. Rev. **D101**, 034512 (2020), 1902.04223
[35] D. Giusti, V. Lubicz, G. Martinelli, F. Sanfilippo, S. Simula, Phys. Rev. **D99**, 114502 (2019), 1901.10462
[36] T. Aoyama et al., Phys. Rept. **887**, 1 (2020), 2006.04822
[37] M. Davier, A. Hoecker, B. Malaescu, Z. Zhang, Eur. Phys. J. C **80**, 241 (2020), 1908.00921
[38] A. Keshavarzi, D. Nomura, T. Teubner, Phys. Rev. **D101**, 014029 (2020), 1911.00367
[39] G. Colangelo, M. Hoferichter, P. Stoffer, JHEP **02**, 006 (2019), 1810.00007
[40] M. Hoferichter, B.L. Hoid, B. Kubis, JHEP **08**, 137 (2019), 1907.01556

Hadronic contributions to the muon $g-2$ in holographic QCD

Josef Leutgeb[1,*], *Jonas* Mager[1,**], and *Anton* Rebhan[1,***]

[1]Institut für Theoretische Physik, Technische Universität Wien, Wiedner Hauptstrasse 8-10, A-1040 Vienna, Austria

Abstract. We discuss the recent progress made in using bottom-up holographic QCD models in calculating hadronic contributions to the anomalous magnetic moment of the muon, in particular the hadronic light-by-light scattering contribution, where holographic QCD naturally satisfies the Melnikov-Vainshtein constraint by an infinite series of axial vector meson contributions.

1 Introduction

The anomaly of the magnetic moment of the muon $a_\mu = (g-2)_\mu/2$ [1] is currently known with an experimental uncertainty of only 41×10^{-11} when the E821/BNL measurement from 2006 [2] is combined with the concordant result obtained in 2021 by the Muon $g-2$ Collaboration at Fermilab [3], corroborating a long-standing discrepancy with the Standard Model prediction, which according to the 2020 White Paper (WP) of the Muon $g-2$ Theory Initiative [4] has a similar estimated error of 43×10^{-11} but a deviation of 4.2 σ in its value. The theoretical error is completely dominated by hadronic contributions, which arise from hadronic vacuum polarization (HVP) and hadronic light-by-light (HLBL) scattering,

$$a_\mu^{\rm HVP,WP} = (6845\pm40)\times10^{-11} \quad (0.6\%\ \text{error}),$$
$$a_\mu^{\rm HLBL,WP} = (92\pm19)\times10^{-11} \quad (20\%\ \text{error}), \quad (1)$$

in (effective) one-loop and two-loop contributions in the muon-photon vertex diagram.

The WP result for the HVP contribution is obtained by a data-driven, so-called dispersive approach, which however has been challenged by a recent lattice QCD result with comparable errors but a central result that is about 3% larger [5], almost eliminating the discrepancy with the experimental result for a_μ, but instead producing a $\sim 3\sigma$ deviation from results based on the R-ratio in $e^+e^-\to$ hadrons.

The WP estimate of the HLBL contribution is only partially determined by data and involves input from hadronic models with larger uncertainties, albeit direct complete lattice evaluations are now gradually bringing down their errors [6, 7].

Holographic QCD (hQCD) is a conjectural approximation to strongly coupled QCD (in its large-N_c limit) based on the more well-established AdS/CFT correspondence [8, 9] and experience with top-down string-theoretical constructions of gauge/gravity dual theories [10–12]. Using a minimal set of parameters, holographic QCD has proved to be capable of good qualitative and often quantitative predictions in hadron physics [13–15] with typical errors of (sometimes less than) 10 to 30%. This is clearly too crude to be of help with the currently most pressing issue of the theory result for $a_\mu^{\rm HVP}$, where the discrepancy between lattice and data-driven approaches is a few percent and where eventually sub-percent accuracy is needed. However, in the case of HLBL contributions, the error is dominated by two contributions, the estimate of the effect of short-distance constraints (SDC) and the contribution of axial-vector mesons, where the WP values are

$$a_\mu^{\rm HLBL,SDC} = (15\pm10)\times10^{-11} \quad (67\%\ \text{error}),$$
$$a_\mu^{\rm HLBL,axials} = (6\pm6)\times10^{-11} \quad (100\%\ \text{error}). \quad (2)$$

It is here that hQCD can provide interesting results, whereas the rather precisely known HVP contribution can be used to assess potential deficiencies of the various hQCD models.

2 Top-down and bottom-up hQCD models

Exact holographic duals are known only in certain limits for highly symmetric theories such as the superconformal $\mathcal{N}=4$ Yang-Mills theory in the limit of large-N_c and infinite 't Hooft coupling g^2N_c. But already in 1998, Witten [10] succeeded in constructing a gauge/gravity dual to the low-energy limit of large-N_c Yang-Mills theory, "top-down" from type-IIA superstring theory, where supersymmetry and conformal symmetry are broken by compactification on one extra dimension beyond the holographic dimension dual to inverse energy. In 2004, Sakai and Sugimoto found a D-brane construction within Witten's model that adds chiral quarks in the fundamental representation [11, 12], thus providing the so far closest dual theory to low-energy QCD in the large-N_c and chiral limit. With having only a mass scale and one dimensionless number (the 't Hooft coupling at the scale of the Kaluza-Klein compactification) the (Witten-)Sakai-Sugimoto model turns out to be remarkably successful

*e-mail: josef.leutgeb@tuwien.ac.at
**e-mail: jonas.mager@tuwien.ac.at
***e-mail: anton.rebhan@tuwien.ac.at

both qualitatively and also quantitatively. Chiral symmetry breaking is a direct consequence of its geometry, while flavor anomalies are naturally realized. However, above the Kaluza-Klein scale $M_{\rm KK} \approx 1$ GeV this model exhibits a different ultra-violet (UV) behavior than QCD, since its actual dual at high energy scales is a 5-dimensional superconformal theory instead of 4-dimensional QCD with asymptotic freedom.

Skipping a string-theoretic top-down derivation, phenomenologically interesting "bottom-up" models of hadron physics were obtained in [13–15] that combine certain features also present in the Sakai-Sugimoto model with a simpler geometry that is asymptotically AdS_5 (and therefore conformal in the UV), breaking conformal symmetry in the infrared (IR) by a hard-wall (HW) cutoff with appropriate boundary conditions or by a soft wall (SW) provided by a nontrivial dilaton [16–24], which has similarities with light-front holographic QCD [25].

A common feature of the various bottom-up and also the top-down Sakai-Sugimoto model is that vector and axial vector mesons as well as pions are described by 5-dimensional Yang-Mills fields $\mathcal{B}^{L,R}_M = \mathcal{B}^V_M \mp \mathcal{B}^A_M$ for the global $U(N_f)_L \times U(N_f)_R$ chiral symmetry of boundary theory with 5-dimensional action

$$S_{\rm YM} \propto \frac{1}{g_5^2}\,\mathrm{tr}\int d^4x \int_0^{z_0} dz\, e^{-\Phi(z)}\sqrt{-g}\,g^{PR}g^{QS} \times\left(\mathcal{F}^{(L)}_{PQ}\mathcal{F}^{(L)}_{RS} + \mathcal{F}^{(R)}_{PQ}\mathcal{F}^{(R)}_{RS}\right), \quad (3)$$

where $P, Q, R, S = 0, \ldots, 3, z$ and $\mathcal{F}_{MN} = \partial_M \mathcal{B}_N - \partial_N \mathcal{B}_M - i[\mathcal{B}_M, \mathcal{B}_N]$ with conformal boundary at $z = 0$, and either a sharp cut-off of AdS_5 at z_0 (the location of the hard wall) or $z_0 = \infty$ (SW) with nontrivial dilaton.

Chiral symmetry is broken either by introducing an extra bifundamental scalar field X dual to quark bilinears $\bar{q}_L q_R$ as in the original HW model of Erlich et al. [13, 14] (HW1), or through different boundary conditions for vector and axial vector fields at z_0 as in the Hirn-Sanz model [15] as well as in the top-down Sakai-Sugimoto model [11, 12].

Vector meson dominance (VMD), in a form necessarily involving an infinite tower of vector mesons, is naturally built in. Photons are described by boundary values of the appropriate combination of the vector gauge fields as those are sourced by quark currents, and they couple through bulk-to-boundary propagators to mesonic degrees of freedom described by normalizable modes.

Of particular importance for the following is that flavor anomalies follow uniquely from 5-dimensional Chern-Simons terms $S^L_{\rm CS} - S^R_{\rm CS}$ with

$$S_{\rm CS} = \frac{N_c}{24\pi^2}\int \mathrm{tr}\left(\mathcal{B}\mathcal{F}^2 - \frac{i}{2}\mathcal{B}^3\mathcal{F} - \frac{1}{10}\mathcal{B}^5\right) \quad (4)$$

in differential-form notation.

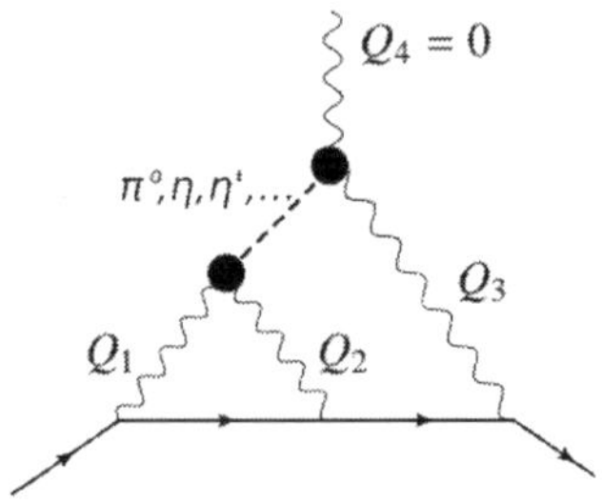

Figure 1. Hadronic light-by-light scattering contribution to a_μ from single meson exchange

3 Anomalous TFFs from hQCD and HLBL contribution to the muon $g-2$

3.1 Holographic pion TFF

The leading HLBL contribution to a_μ comes from the π^0 exchange diagram shown in fig. 1 due to the anomalous coupling of the pion to two photons. This involves both a singly-virtual transition form factor (TFF) in the upper vertex and a doubly-virtual one in the interior of the diagram. The holographic prediction is given by

$$F_{\pi^0\gamma^*\gamma^*}(Q_1^2, Q_2^2) = -\frac{N_c}{12\pi^2 f_\pi}\int_0^{z_0} dz\, \mathcal{J}(Q_1, z)\mathcal{J}(Q_2, z)\Psi(z), \quad (5)$$

where $\mathcal{J}(Q, z)$ is the bulk-to-boundary propagator of a photon with virtuality $Q^2 = -q^2$ and a holographic pion profile function $\Psi(s)$. This has been studied by Grigoryan and Radyushkin in [26–28], who noticed that hQCD models with asymptotic AdS_5 geometry reproduce the asymptotic momentum dependence obtained by Brodsky and Lepage [29–31],

$$F_{\pi^0\gamma^*\gamma^*}(Q_1^2, Q_2^2) \to \frac{2f_\pi}{Q^2}\left[\frac{1}{w^2} - \frac{1-w^2}{2w^3}\ln\frac{1+w}{1-w}\right] \quad (6)$$

with $Q^2 = \frac{1}{2}(Q_1^2 + Q_2^2) \to \infty$, $w = (Q_1^2 - Q_2^2)/(Q_1^2 + Q_2^2)$, which is not achieved by conventional VMD models.

With certain simplifications the holographic results for the pion TFF have been employed in ref. [32, 33] to evaluate the a_μ contribution given by the diagram in fig. 1. In ref. [34] we have carried out a complete evaluation and found good agreement with the data-driven (dispersive) result, which is bracketed by the HW1 and HW2 results when these models are fitted to f_π and m_ρ. With such a fit, the HW2 model, which has one parameter less than the chiral HW1 model, actually undershoots the asymptotic limit of (6) by 38% (fitting the asymptotic limit would instead lead to an overweight ρ meson with mass of 987 MeV). Its prediction of $a_\mu^{\pi^0} = 56.9 \times 10^{-11}$ is correspondingly on the low side. The HW1 model, which satisfies (6) exactly, predicts $a_\mu^{\pi^0} = 65.2 \times 10^{-11}$, which is consistent with, but somewhat larger than, the WP value $62.6^{+3.0}_{-2.5} \times 10^{-11}$.

However, at not too large, phenomenologically relevant energy scales, the asymptotic behavior of the TFF is

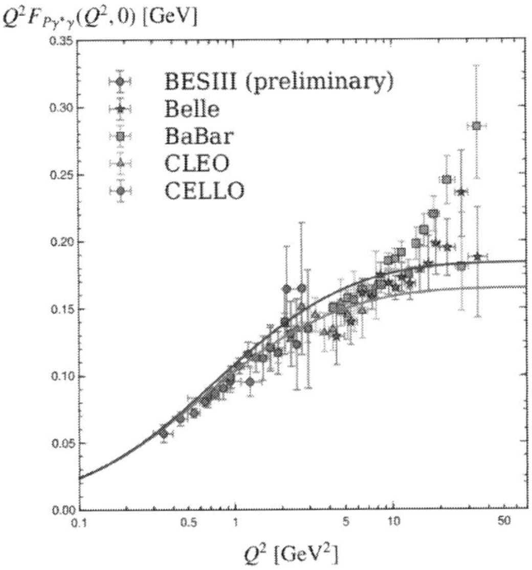

Figure 2. Holographic results for the single virtual TFF $Q^2F(Q^2, 0)$ for π^0, plotted on top of experimental data as compiled in Fig. 53 of Ref. [4] for $g_5 = 2\pi$ (OPE fit, blue) and the reduced value (red) corresponding to a fit of F_ρ. (Taken from [41])

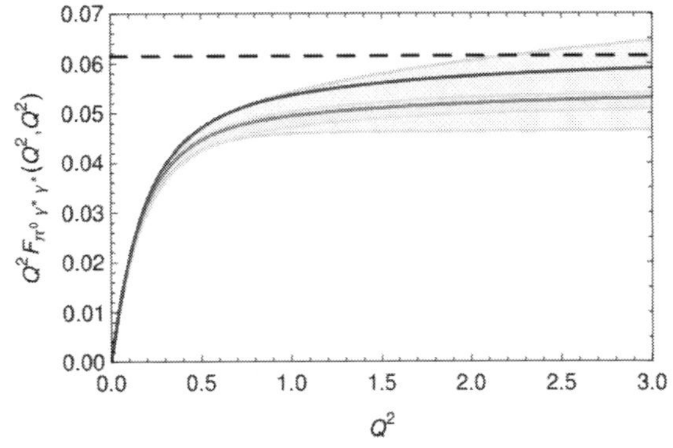

Figure 3. Holographic results for the doubly virtual $F_{\pi^0\gamma^*\gamma^*}$ compared to the dispersive result of Ref. [39] (green band) and the lattice result of Ref. [40] (yellow band); the OPE limit is indicated by the dashed horizontal line. The blue line corresponds to $g_5 = 2\pi$ (OPE fit) and the red one to the reduced value g_5 (F_ρ-fit). (Taken from [41])

modified by gluonic corrections of the order of 10% [35, 36], while hQCD models with a simple AdS$_5$ background approach the asymptotic limit somewhat too quickly. Similar corrections but with opposite sign appear in the vector correlator appearing in HVP. Indeed, the HW1 model with complete UV fit underestimates HVP. Reducing g_5^2 by 10% (which is achieved by fitting instead the decay constant of the ρ meson) brings the HVP in line with the dispersive result (to within 5%) [37] and also makes the HLBL result $a_\mu^{\pi^0}$ completely coincident with the central WP value [38].

In fig. 2 the HW1 result (with finite quark masses) for the singly virtual pion TFF is compared with experimental data, where the result with reduced g_5^2 is seen to agree somewhat better at the relevant energy scale $Q^2 \lesssim 10\,\mathrm{GeV}^2$; in fig. 3 the doubly virtual result for $Q_1^2 = Q_2^2$ is compared with the dispersive result of ref. [39] and the lattice result of Ref. [40]. Here the HW1 result with reduced coupling coincides with the central result of the dispersive approach within line thickness and it also happens to be in the center of the lattice result.

3.2 Axial vector meson TFF

In hQCD, the coupling of axial vector mesons to photons is determined by the same action (4) that gives rise to the anomalous TFFs of the pseudoscalars. There is an infinite tower of axial vector mesons and their decay amplitude has the form [42, 43]

$$\mathcal{M}_{\mathcal{A}\gamma^*\gamma^*} = i\frac{N_c}{4\pi^2}\mathrm{tr}(Q^2 t^a)\,\epsilon_{(1)}^\mu \epsilon_{(2)}^\nu \epsilon_{\mathcal{A}}^{*\rho}\epsilon_{\mu\nu\rho\sigma} \times \left[q_{(2)}^\sigma Q_1^2 A_n^a(Q_1^2, Q_2^2) - q_{(1)}^\sigma Q_2^2 A_n^a(Q_2^2, Q_1^2)\right], \quad (7)$$

involving an asymmetric structure function

$$A_n(Q_1^2, Q_2^2) = \frac{2g_5}{Q_1^2}\int_0^{z_0} dz \left[\frac{d}{dz}\mathcal{J}(Q_1, z)\right]\mathcal{J}(Q_2, z)\,\psi_n^A(z) \quad (8)$$

where $\psi_n^A(z)$ is the holographic wave function of the n-th axial vector meson. The most general decay amplitude of axial vector mesons would permit a second asymmetric structure function [44–46], but this does not appear in the simplest versions of hQCD considered here.

The holographic result satisfies the Landau-Yang theorem [47, 48], which forbids the decay of an axial vector meson in two real photons, due to the fact that $\mathcal{J}'(Q, z) = 0$ for $Q^2 = 0$.

The asymptotic form of $A_n(Q_1^2, Q_2^2)$ for large virtualities is given by [42]

$$A_n(Q_1^2, Q_2^2) \to \frac{12\pi^2 F_n^A}{N_c Q^4}\frac{1}{w^4}\left[w(3-2w) + \frac{1}{2}(w+3)(1-w)\ln\frac{1-w}{1+w}\right], \quad (9)$$

$w = (Q_1^2 - Q_2^2)/(Q_1^2 + Q_2^2)$, which agrees with the independent pQCD result obtained recently in [49].

At moderate virtualities, the shape of the singly virtual axial TFFs turns out to be consistent with the dipole fit of data obtained by the L3 experiment for $f_1(1285)$ [51], as shown in fig. 4. The overall magnitude, which for HW1 and HW2 models is found as 21.04 and 16.63 GeV^{-2} is in roughly the right ballpark compared to experimental data, albeit somewhat too high for HW1 which overestimates the equivalent photon widths of $f_1(1285)$ and $f_1' = f_1(1420)$ [51, 52], but consistent with the phenomenological value $A(0, 0)_{a_1(1230)} = 19.3(5.0)\,\mathrm{GeV}^{-2}$ obtained in [45, 53].

The results from the L3 experiment have been used by Pauk and Vanderhaeghen [50] to estimate the contributions of f_1 and f_1' to a_μ, assuming a symmetric factorized dipole moment $A^{\mathrm{PV}}(Q_1^2, Q_2^2)/A(0, 0) = (1 + Q_1^2/\Lambda_D^2)^{-2}(1 +$

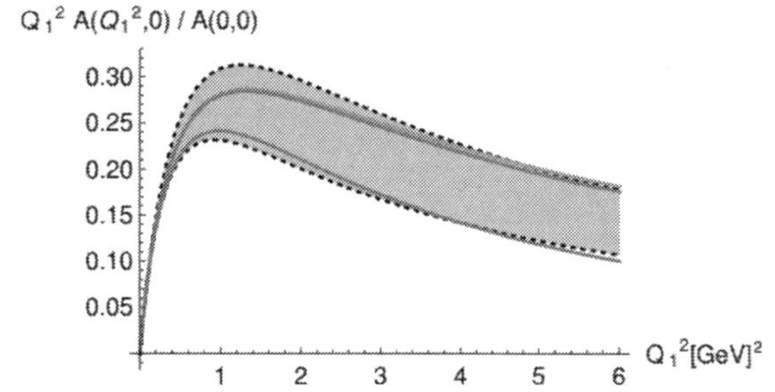

Figure 4. Single-virtual axial vector TFF from holographic models (SS: blue, HW1: orange, HW2: red) compared with dipole fit of L3 data for $f_1(1285)$ (grey band). The parameters of all models are fixed by matching f_π and m_ρ. The results for HW1 and HW2 almost coincide, with HW2 at most a line thickness above HW1. (Taken from [42])

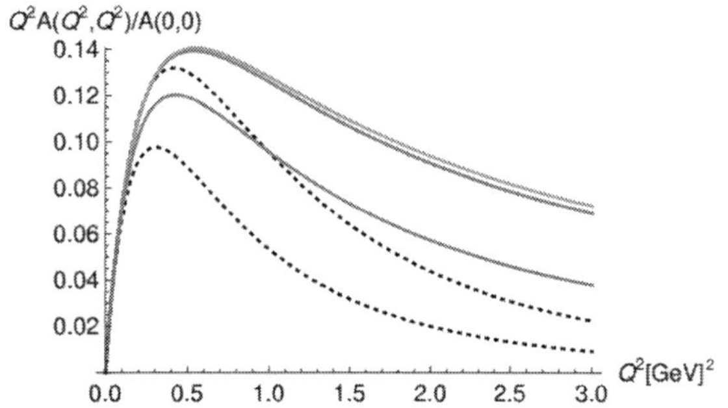

Figure 5. Double-virtual axial vector TFF for $Q_1^2 = Q_2^2 = Q^2$ from holographic models (SS: blue, HW1: orange, HW2: red). The black dashed lines denote the extrapolation of L3 data with a dipole model for each virtuality as used in the calculation of $a_\mu^{f_1}$ in Ref. [50]. (Taken from [42])

$Q_2^2/\Lambda_D^2)^{-2}$. This drops too quickly in the doubly virtual case to match the short-distance behavior required by pQCD. In fig. 5 the hQCD results are shown and compared with this simple ansatz, exhibiting a considerable difference.

3.3 Melnikov-Vainshtein SDC

In [42, 43] it has been shown moreover that in hQCD the infinite tower of axial vector mesons is responsible for the correct implementation of the Melnikov-Vainshtein SDC [54] for the HLBL scattering amplitude, which is a consequence of the nonrenormalization theorem for the chiral anomaly. In terms of the tensor basis used in [55] it reads

$$\lim_{Q_3\to\infty}\lim_{Q\to\infty} Q_3^2 Q^2 \bar{\Pi}_1(Q,Q,Q_3) = -\frac{2}{3\pi^2}. \tag{10}$$

Clearly, each single meson exchange contribution gives a vanishing result in this limit, because the propagator in fig. 1 goes like $1/Q_3^2$ and the two form factors go like $1/Q^2$ and $1/Q_3^2$.

In [54] Melnikov and Vainshtein proposed to estimate the impact of the MV-SDC by replacing the external TFF by its constant on-shell value. With current input data, this would lead to an increase of almost 40% of the pseudoscalar contribution $a_\mu^{\pi^0,\eta,\eta'}$ by 38×10^{-11}. However, the MV-SDC can also be satisfied by having an infinite tower of single-meson contributions. In Ref. [56, 57] Colangelo et al. showed that a Regge model for a tower of excited pseudoscalars can be constructed that saturates the MV-SDC with a contribution $\Delta a_\mu^{\rm PS} = 13(6)\times10^{-11}$ when using the available experimental constraints for the photon decay amplitudes of excited pseudoscalars. However, as also noted in [57], excited pseudoscalar states decouple in the chiral large-N_c limit, which is not the case for the infinite tower of axial vector mesons. Indeed, in holographic QCD it is the latter which are responsible for the implementation of the MV-SDC [42, 43]. This is illustrated in fig. 6 for the simple HW2 model, which permits a closed form solution for the full series, for $\Pi_1(Q,Q,Q_3)$ with large $Q = 50$ GeV and increasing $Q_3 \ll Q$: while each axial vector meson mode gives an asymptotically vanishing contribution, their infinite series satisfies the MV-SDC constraint.

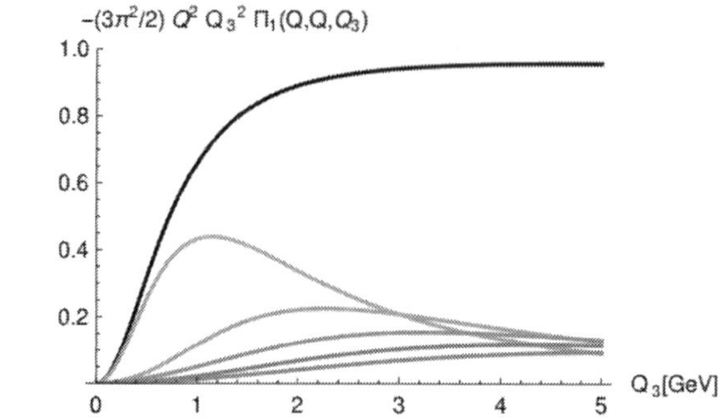

Figure 6. Axial-vector contribution to $Q_3^2 Q^2 \bar{\Pi}_1(Q,Q,Q_3)$ as a function of Q_3 at $Q = 50$ GeV in the HW2 model (with $g_5 = 2\pi$ so that the SDC is satisfied to 100%). The black line corresponds to the infinite sum over the tower of axial vector mesons, and the colored lines give the contributions of the 1st to 5th lightest axial vector mesons. (Taken from [42])

In [38] two of us have shown that also in the HW1 model with massive quarks the infinite tower of axial vector mesons is responsible for the realization of the MV-SDC. The infinite tower of excited pseudoscalars, which does not decouple from the anomaly relation away from the chiral limit, contributes only subleading terms $\propto \ln(Q_3^2)/Q_3^4 Q^2$ to (10).

The hQCD models generally satisfy the MV-SDC to the same level as they satisfy the SDCs on the TFFs (see the remarks in sect. 3.1.) In the limit where all Q_i become large simultaneously, another SDC can be derived [54, 58] for $\bar{\Pi}$, $\lim_{Q\to\infty} Q^4\bar{\Pi}_1(Q,Q,Q) = -4/9\pi^2$. In hQCD models with correct short-distance limit of the TFFs, the coefficient on the right-hand side comes out 19% smaller [38, 43].

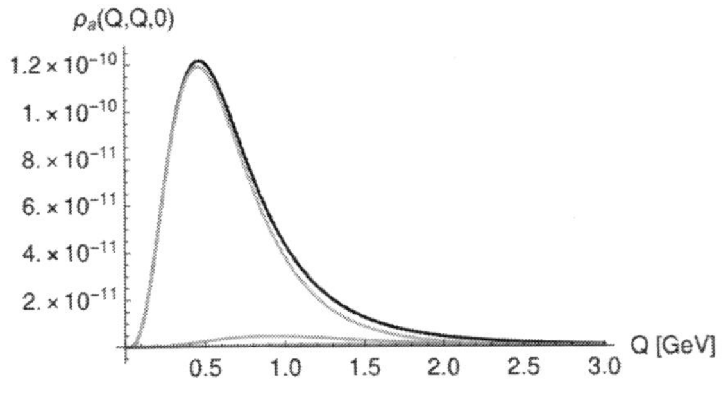

Figure 7. The integrand $\rho_a(Q_1, Q_2, \tau)$ in (11) in units of GeV^{-2} for $Q_1 = Q_2 = Q$ and $\tau = 0$ in the case of the HW2 model. The black line is the result from the infinite sum over the tower of axial vector mesons, the colored lines give the contributions of the 1st to 3rd lightest axial vector meson multiplets. (Taken from [42])

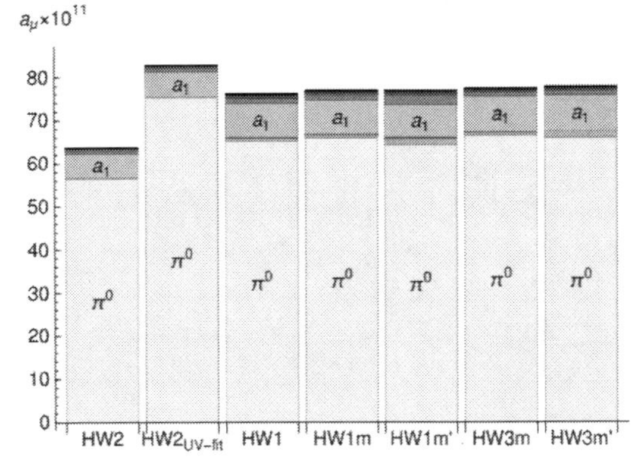

Figure 8. Bar chart of the individual contributions to a_μ from the isotriplet sector in the Hirn-Sanz model fitted to IR (HW2) and UV data (HW2$_{\text{UV-fit}}$), the chiral HW1 model and various extensions to the massive case, with excited modes given by increasingly darker colors, blue for the π^0's, red for the a_1's. (Taken from [38])

3.4 a_μ contributions

Fig. 7 shows the individual contributions of axial vector meson excitations to the integrand in

$$a_\mu^{\text{AV}} = \int_0^\infty dQ_1 \int_0^\infty dQ_2 \int_{-1}^1 d\tau\, \rho_a(Q_1, Q_2, \tau) \qquad (11)$$

for $Q_1 = Q_2$ and $\tau = 0$. The ground-state axial vector meson mode clearly dominates, but the infinite sum of excited modes does contribute non-negligibly. For chiral models, this is shown quantitatively in table 2. In the Sakai-Sugimoto model, where the SDCs are not satisfied at all, excited modes contribute only a few percent. In the HW2 model, which satisfies the SDCs qualitatively, but quantitatively only at the level of 62% when f_π and m_ρ are fitted to physical values, the excited modes add +25% to the contribution of the ground-state mode ($j = 1$). This percentage increases to about 30% in the HW1 model, where the SDCs are matched exactly. Also shown in table 2 are two modified versions of the HW1 model with g_5^2 reduced by 10% and 15%, which can be viewed as taking into account next-to-leading order corrections at large but still physically relevant energy scales. Also in these models, the excited modes add around 30%, if not more, to $a_\mu^{\text{AV}}(j = 1)$.

While being completely absent in the chiral HW2 model, excited pseudoscalars, which decouple from the axial current in the chiral limit, always contribute to a_μ in the HW1 model and its variants. This has been studied in [38] in the flavor-symmetric case with and without modified scaling dimension of the bifundamental scalar and also for different boundary conditions on the hard wall. The various models show some differences in the composition of the contributions to a_μ, but hardly deviate from the sum total in the chiral model (where the physical pion mass is inserted manually in the pion propagator).

In the flavor-U(3)-symmetric case the total axial vector sector is four times the contribution of the a_1 meson. With a 10% reduction of g_5^2 and correspondingly the level of saturation of the SDCs, the massive HW models together yield $a_\mu^{\text{AV}} = 4a_\mu^{a_1} = 39.5(1.6) \times 10^{-11}$, where approximately 58% are due to longitudinal contributions entering $\bar{\Pi}_1$. The contribution of excited pseudoscalars varies somewhat more, as can be seen from fig. 8. In models where their contribution is larger, the ground-state yields less, but we have $a_\mu^{\text{PS}^*} \gtrsim 3.3 \times 10^{-11}$ (again with U(3) symmetry and with 10% reduction of g_5^2). Even without the latter contribution, the holographic result is somewhat higher than the WP estimate for the combined effect of axial vector mesons and SDCs, $a_\mu^{\text{axials+SDC(WP)}} = 21(16) \times 10^{-11}$.

3.5 Katz-Schwartz model: HW1 with solved U(1)$_A$ problem

In the above estimates obtained from flavor-U(3)-symmetric HW models, the effects of the large strange quark mass and the anomalous breaking of U(1)$_A$ symmetry has been ignored (in [42, 43] the contribution of η and η' mesons, where U(3) symmetry clearly is no good approximation, has been estimated by manually setting their masses and decay constants to physical values).

In [41] we have recently considered the minimal extension[1] of the HW1 proposed by Katz and Schwartz [60, 61] to implement the U(1)$_A$ anomaly through an additional complex scalar Y whose magnitude and phase are dual to the gluon operators $\alpha_s G_{\mu\nu}^2$ and $\alpha_s G\tilde{G}$. There a coupling term $\kappa Y^{N_f} \det(X) \subset \mathcal{L}$ in the 5-dimensional theory accounts for U(1)$_A$ anomalous Ward identities. It turns out that all results are weakly dependent on the new coupling κ when $\kappa \gg 1$. The only new free parameter is the value of the gluon condensate Ξ which can be prescribed in the choice of the background field $\langle Y \rangle = C + \Xi z^4$, while C is fixed by OPE and the U(1)$_A$ anomaly to $C \propto \alpha_s$. The latter

[1] In [59] an extension of the HW1 model was considered which in contrast to the minimal models leads to nonvanishing HLBL contributions of scalar mesons. While results were obtained that are consistent with previous estimates of these contributions, the scalar TFFs obtained therein do not agree with pQCD SDC constraints [49].

Table 1. Axial vector contributions in chiral HW models with different levels of saturation of SDCs. In the chiral, flavor U(3)-symmetric case we have simply $a_\mu^{AV} = a_\mu^{a_1+f_1+f_1'} = 4a_\mu^{a_1}$. Also given are the masses m_{a_1} and the photon coupling amplitudes $A_1(0,0)$ of the ground-state axial vector mesons ($j = 1$).

chiral model	m_{a_1}	$A_1(0,0)$	SDC	$a_\mu^{AV}(j=1)$	$(j \leq 2)$	$(j \leq 3)$	$(j \leq 4)$	$(j \leq 5)$	a_μ^{AV}(all)
HW1	1.3755	21.04	100%	31.4	36.2	37.9	39.1	39.6	41×10^{-11}
HW1-	1.295	19.65	90%	28.7	34.3	36.0	37.1	37.5	38×10^{-11}
HW1--	1.255	18.90	85%	27.3	33.2	34.9	35.9	36.3	37×10^{-11}
HW2	1.235	16.63	62%	23.0	26.2	27.4	27.9	28.2	29×10^{-11}
SS	1.1865	15.93	0%	13.8	14.5	14.7	14.8	14.8	15×10^{-11}

is modeled by $\alpha_s \to 1/\beta_0 \ln(\Lambda_{QCD} z)$, $\Lambda_{QCD} \to z_0^{-1}$. This model implements a Witten-Veneziano mechanism for the mass of isosinglet pseudoscalars, which are augmented by a pseudoscalar glueball mode mixing with the η and η' mesons.

In [60], Katz and Schwartz have found that their model can account for the masses of η and η' mesons with deviations of about 10%. By turning on the gluon condensate parameter Ξ, we have found [41] that this can be improved to the percent level.

In the case of axial vector mesons, this model is somewhat less successful: the masses of f_1 and f_1' are raised too much compared to their experimental values, and also the experimental values of their mixing is not reproduced. Nevertheless, this provides a first glimpse of what a more complete hQCD model can give.

The results that we have obtained with the Katz-Schwartz model augmented by a gluon condensate are given in table 2. In the isotriplet (pion and a_1) sector, the results are essentially unchanged, with the pion contribution matching perfectly the dispersive result, in particular when g_5 is fitted to match F_ρ, corresponding to a 90% level of saturation of the OPE constraints. The contributions from η and η' mesons are then also completely in line with the WP estimates. Excited pseudoscalars (including a pseudoscalar glueball mode) contribute another 1.6×10^{-11}, so that the total contribution of pseudoscalars is at the upper bound of the WP estimate. The combined contribution of ground-state axial vector mesons equals around 3.5 times the a_1 contribution, which is a moderate reduction of the factor 4 in the flavor-U(3) symmetric case. Somewhat surprisingly, the contribution of excited f_1 and f_1' axial vector mesons is more strongly reduced, so that the "best guess" for the axials + LSDC contributions in this model turns out to be $a_\mu^{AV+LSDC} = 30.5 \times 10^{-11}$, and $a_\mu^{AV+PS^*+LSDC} = 32 \times 10^{-11}$ when excited pseudoscalars are included. Both results are somewhat above but now safely within the estimated error of the WP estimate $a_\mu^{\text{axials+SDC(WP)}} = 21(16) \times 10^{-11}$.

We should like to point out, however, that there are holographic QCD models that are closer to a string-theoretic top-down construction such as the typically more involved models of Ref. [62–64]. It would be interesting to see if they can achieve a better fit of masses and mixing angles of f_1, f_1' mesons, and what they would predict for their contribution to a_μ.

Table 2. Summary of the results [41] for the different contributions to a_μ in the massive HW model due to Katz and Schwartz (but augmented by a nonzero gluon condensate (v1)) in comparison with the White Paper [4] values. The mass of π^0 is fixed as input, the masses of η and η' are matched within 0.8 and 2.4%, while the axial vector mesons a_1 and f_1 are between 4% and 15% too heavy, f_1' by 28%.

$a_\mu^{\dots} \times 10^{11}$	v1(OPE fit)	v1(F_ρ-fit)	WP
π^0	66.1	63.4	$62.6^{+3.0}_{-2.5}$
η	19.3	17.6	16.3(1.4)
η'	16.9	14.9	14.5(1.9)
G/η''	0.2	0.2	
$\sum_{PS^*}$	1.6	1.4	
PS poles	104	97.5	93.8(4.0)
a_1	7.8	7.1	
$f_1 + f_1'$	20.0	17.9	
$\sum_{a_1^*}$	2.2	2.4	
$\sum_{f_1^{(\prime)*}}$	3.6	3.0	
AV+LSDC	33.7	30.5	21(16)
total	138	128	115(16.5)

Acknowledgments

We would like to thank Luigi Cappiello, Gilberto Colangelo, Giancarlo D'Ambrosio, Martin Hoferichter, Elias Kiritsis, and Pablo Sanchez-Puertas for useful discussions. J. L. and J. M. have been supported by the Austrian Science Fund FWF, project no. P33655, and by the FWF doctoral program Particles & Interactions, project no. W1252-N27.

References

[1] F. Jegerlehner, Springer Tracts Mod. Phys. **274**, pp.1 (2017)

[2] G.W. Bennett et al. (Muon g-2), Phys. Rev. D **73**, 072003 (2006), `hep-ex/0602035`

[3] B. Abi et al. (Muon g-2), Phys. Rev. Lett. **126**, 141801 (2021), `2104.03281`

[4] T. Aoyama et al., Phys. Rept. **887**, 1 (2020), `2006.04822`

[5] S. Borsanyi et al., Nature **593**, 51 (2021), `2002.12347`

[6] T. Blum, N. Christ, M. Hayakawa, T. Izubuchi, L. Jin, C. Jung, C. Lehner, Phys. Rev. Lett. **124**, 132002 (2020), 1911.08123

[7] E.H. Chao, R.J. Hudspith, A. Gérardin, J.R. Green, H.B. Meyer, K. Ottnad, Eur. Phys. J. C **81**, 651 (2021), 2104.02632

[8] J.M. Maldacena, Int.J.Theor.Phys. **38**, 1113 (1999), hep-th/9711200

[9] O. Aharony, S.S. Gubser, J.M. Maldacena, H. Ooguri, Y. Oz, Phys.Rept. **323**, 183 (2000), hep-th/9905111

[10] E. Witten, Adv.Theor.Math.Phys. **2**, 505 (1998), hep-th/9803131

[11] T. Sakai, S. Sugimoto, Prog.Theor.Phys. **113**, 843 (2005), hep-th/0412141

[12] T. Sakai, S. Sugimoto, Prog.Theor.Phys. **114**, 1083 (2005), hep-th/0507073

[13] J. Erlich, E. Katz, D.T. Son, M.A. Stephanov, Phys. Rev. Lett. **95**, 261602 (2005), hep-ph/0501128

[14] L. Da Rold, A. Pomarol, Nucl. Phys. **B721**, 79 (2005), hep-ph/0501218

[15] J. Hirn, V. Sanz, JHEP **12**, 030 (2005), hep-ph/0507049

[16] K. Ghoroku, N. Maru, M. Tachibana, M. Yahiro, Phys. Lett. B **633**, 602 (2006), hep-ph/0510334

[17] A. Karch, E. Katz, D.T. Son, M.A. Stephanov, Phys. Rev. **D74**, 015005 (2006), hep-ph/0602229

[18] H.J. Kwee, R.F. Lebed, JHEP **01**, 027 (2008), 0708.4054

[19] U. Gürsoy, E. Kiritsis, JHEP **02**, 032 (2008), 0707.1324

[20] U. Gürsoy, E. Kiritsis, F. Nitti, JHEP **02**, 019 (2008), 0707.1349

[21] P. Colangelo, F. De Fazio, F. Giannuzzi, F. Jugeau, S. Nicotri, Phys. Rev. D **78**, 055009 (2008), 0807.1054

[22] T. Gherghetta, J.I. Kapusta, T.M. Kelley, Phys. Rev. D **79**, 076003 (2009), 0902.1998

[23] T. Branz, T. Gutsche, V.E. Lyubovitskij, I. Schmidt, A. Vega, Phys. Rev. D **82**, 074022 (2010), 1008.0268

[24] P. Colangelo, F. De Fazio, J.J. Sanz-Cillero, F. Giannuzzi, S. Nicotri, Phys. Rev. **D85**, 035013 (2012), 1108.5945

[25] S.J. Brodsky, G.F. de Teramond, H.G. Dosch, J. Erlich, Phys. Rept. **584**, 1 (2015), 1407.8131

[26] H.R. Grigoryan, A.V. Radyushkin, Phys. Rev. **D76**, 115007 (2007), 0709.0500

[27] H.R. Grigoryan, A.V. Radyushkin, Phys. Rev. **D77**, 115024 (2008), 0803.1143

[28] H.R. Grigoryan, A.V. Radyushkin, Phys. Rev. **D78**, 115008 (2008), 0808.1243

[29] S.J. Brodsky, G.P. Lepage, Phys. Rev. **D24**, 1808 (1981)

[30] G.P. Lepage, S.J. Brodsky, Phys. Lett. **87B**, 359 (1979)

[31] G.P. Lepage, S.J. Brodsky, Phys. Rev. **D22**, 2157 (1980)

[32] D.K. Hong, D. Kim, Phys. Lett. **B680**, 480 (2009), 0904.4042

[33] L. Cappiello, O. Cata, G. D'Ambrosio, Phys. Rev. **D83**, 093006 (2011), 1009.1161

[34] J. Leutgeb, J. Mager, A. Rebhan, Phys. Rev. **D100**, 094038 (2019), erratum Phys. Rev. D 104 (2021) 059903, 1906.11795

[35] B. Melic, D. Mueller, K. Passek-Kumericki, Phys. Rev. **D68**, 014013 (2003), hep-ph/0212346

[36] J. Bijnens, N. Hermansson-Truedsson, L. Laub, A. Rodríguez-Sánchez, JHEP **04**, 240 (2021), 2101.09169

[37] J. Leutgeb, A. Rebhan, M. Stadlbauer, Phys. Rev. D **105**, 094032 (2022), 2203.16508

[38] J. Leutgeb, A. Rebhan, Phys. Rev. D **104**, 094017 (2021), 2108.12345

[39] M. Hoferichter, B.L. Hoid, B. Kubis, S. Leupold, S.P. Schneider, JHEP **10**, 141 (2018), 1808.04823

[40] A. Gérardin, H.B. Meyer, A. Nyffeler, Phys. Rev. D **100**, 034520 (2019), 1903.09471

[41] J. Leutgeb, J. Mager, A. Rebhan (2022), 2211.16562

[42] J. Leutgeb, A. Rebhan, Phys. Rev. D **101**, 114015 (2020), 1912.01596

[43] L. Cappiello, O. Catà, G. D'Ambrosio, D. Greynat, A. Iyer, Phys. Rev. D **102**, 016009 (2020), 1912.02779

[44] V. Pascalutsa, V. Pauk, M. Vanderhaeghen, Phys. Rev. **D85**, 116001 (2012), 1204.0740

[45] P. Roig, P. Sanchez-Puertas, Phys. Rev. D **101**, 074019 (2020), 1910.02881

[46] M. Zanke, M. Hoferichter, B. Kubis, JHEP **07**, 106 (2021), 2103.09829

[47] L.D. Landau, Dokl. Akad. Nauk Ser. Fiz. **60**, 207 (1948)

[48] C.N. Yang, Phys. Rev. **77**, 242 (1950)

[49] M. Hoferichter, P. Stoffer, JHEP **05**, 159 (2020), 2004.06127

[50] V. Pauk, M. Vanderhaeghen, Eur. Phys. J. **C74**, 3008 (2014), 1401.0832

[51] P. Achard et al. (L3), Phys. Lett. **B526**, 269 (2002), hep-ex/0110073

[52] P. Achard et al. (L3), JHEP **03**, 018 (2007)

[53] P. Masjuan, P. Roig, P. Sanchez-Puertas (2020), 2005.11761

[54] K. Melnikov, A. Vainshtein, Phys. Rev. **D70**, 113006 (2004), hep-ph/0312226

[55] G. Colangelo, M. Hoferichter, M. Procura, P. Stoffer, JHEP **09**, 074 (2015), 1506.01386

[56] G. Colangelo, F. Hagelstein, M. Hoferichter, L. Laub, P. Stoffer, Phys. Rev. D **101**, 051501 (2020), 1910.11881

[57] G. Colangelo, F. Hagelstein, M. Hoferichter, L. Laub, P. Stoffer, JHEP **03**, 101 (2020), 1910.13432

[58] J. Bijnens, N. Hermansson-Truedsson, A. Rodríguez-Sánchez, Phys. Lett. B **798**, 134994 (2019), 1908.03331

[59] L. Cappiello, O. Catà, G. D'Ambrosio, Phys. Rev. D **105**, 056020 (2022), 2110.05962

[60] E. Katz, M.D. Schwartz, JHEP **08**, 077 (2007), 0705.0534

[61] T. Schäfer, Phys. Rev. D **77**, 126010 (2008), 0711.0236

[62] R. Casero, E. Kiritsis, A. Paredes, Nucl. Phys. B **787**, 98 (2007), hep-th/0702155

[63] D. Areán, I. Iatrakis, M. Järvinen, E. Kiritsis, JHEP **1311**, 068 (2013), 1309.2286

[64] F. Giannuzzi, S. Nicotri, Phys. Rev. D **104**, 014021 (2021), 2105.00923

Flavor anomalies meet the LHC

Syuhei Iguro[1,2,*,**]

[1] *Institute for Theoretical Particle Physics (TTP), Karlsruhe Institute of Technology (KIT), Engesserstraße 7, 76131 Karlsruhe, Germany*
[2] *Institute for Astroparticle Physics (IAP), Karlsruhe Institute of Technology (KIT), Hermann-von-Helmholtz-Platz 1, 76344 Eggenstein-Leopoldshafen, Germany*

Abstract. Currently there are several discrepancies between the standard model (SM) prediction and experimental value in the flavor observables, *e.g.* R_{D^*}, $\mathrm{BR}(\bar{B} \to D^+K^-)$, $R_{K^{(*)}}$ and muon anomalous magnetic moment. If we interpret the deviation as a hint for the physics beyond the SM, $\mathcal{O}(1)$ TeV new particles are often implied. Therefore it is natural to search those new particles at the large hadron colliders. In this note we interplay the flavor anomalies and collider physics.

1 Introduction

In recent decades, we have performed high precision verification of the Standard Model (SM) and the long awaited scalar (Higgs boson) was found in 2012. In the meantime many physical observables have been measured with the high accuracy and their SM predictions have been also sharpened. As a result we can test not only the SM but also new physics (NP) beyond the SM, by comparing the model predictions with the experimental results. Most of the results suggest that the SM describes our nature very well, while we also find some measurements deviated from the SM predictions by more than $3-4\,\sigma$. Those flavor anomalies are one of the hot topics of the particle physics [1]. In this note we discuss the following flavor anomalies in $R_{D^{(*)}} = BR(\bar{B} \to D^{(*)}\tau\bar{\nu})/BR(\bar{B} \to D^{(*)}\ell\bar{\nu})$ where $\ell = e,\, \mu$, non leptonic two body B decays ($b \to cu\bar{q}$ puzzle), and muon anomalous magnetic moment (muon $g-2$). There are discrepancies in $b \to s\mu\bar{\mu}$, however, the implied NP scale is $\mathcal{O}(10)$ TeV. Therefore collider physics is less relevant to $b \to s\mu\bar{\mu}$, and is not discussed here.[1]

The semi-tauonic B-meson decays have been intriguing processes to measure the lepton flavor universality (LFU) which is one of the most important prediction of the SM. The ratio $R_{D^{(*)}}$ has been measured by B-factories for more than 10 years and the world average deviates from the SM prediction [5]. Very recently the LHCb collaboration reported the new result which is again off from the SM prediction [6]. The current status is given as

$$R_D^{\mathrm{WA}} = 0.358 \pm 0.028\,, \quad R_{D^*}^{\mathrm{WA}} = 0.285 \pm 0.013\,, \tag{1}$$

$$R_D^{\mathrm{SM}} = 0.248 \pm 0.001, \quad R_{D^*}^{\mathrm{SM}} = 0.289 \pm 0.004, \tag{2}$$

with the experimental correlation of -0.27 and the current significance is about 4σ [7]. Within the SM the tree level W boson exchange describes the transition and the required size of the shift in $R_{D^{(*)}}$ is 10-20%. Because of the size of the required shift, we can safely focus on the operators up to dimension 6 and then the variety of the possible new mediator particles is limited. Currently there are only two categories of the scenarios available. One is a charged Higgs (H^-) and the other is a leptoquark (X). In this proceedings, we will discuss the LHC phenomenology of the those scenarios. Since the 4-fermi interaction to enhance the $R_{D^{(*)}}$ automatically generates $\tau\nu$ final state we focus on the $\tau\nu(+b)$ signature. Although we focus on the collider phenomenology in this note, it is known that polarization observables of the $\bar{B} \to D^{(*)}\tau\bar{\nu}$ transition have a potential to discriminate the those new physics scenarios [8].

In addition to the semileptonic decays, there are coherent deviations between the SM prediction based on the QCD factorization and experimental measurements in hadronic two body decays [9]. It is worthwhile to mention that three independent experiments agree well within the uncertainty [10]. Those deviations are known in so-called color allowed decays of $\bar{B} \to DP$ and $\bar{B} \to D^*P$ transitions where P denotes a pseudoscalar meson. Those decays are also mediated by the tree level W boson exchange, and the comparison between the experimental average and the theoretical predictions are given as [11]

$$BR(\bar{B}^0 \to D^+K^-)^{\mathrm{exp}}_{\mathrm{SM}} = \begin{cases} (1.86 \pm 0.20) \times 10^{-4}\,, \\ (3.26 \pm 0.15) \times 10^{-4}\,, \end{cases} \tag{3}$$

$$BR(\bar{B}^0_s \to D^+_s\pi^-)^{\mathrm{exp}}_{\mathrm{SM}} = \begin{cases} (3.00 \pm 0.23) \times 10^{-3}\,, \\ (4.09 \pm 0.21) \times 10^{-3}\,, \end{cases} \tag{4}$$

$$BR(\bar{B}^0 \to D^{*+}K^-)^{\mathrm{exp}}_{\mathrm{SM}} = \begin{cases} (2.12 \pm 0.15) \times 10^{-4}\,, \\ (3.03 \pm 0.15) \times 10^{-4}\,, \end{cases} \tag{5}$$

*e-mail: igurosyuhei@gmail.com
** preprint number: TTP22-066, P3H-22-107
[1] For the interested readers, see, Refs. [2–4] which discuss the collider phenomenology.

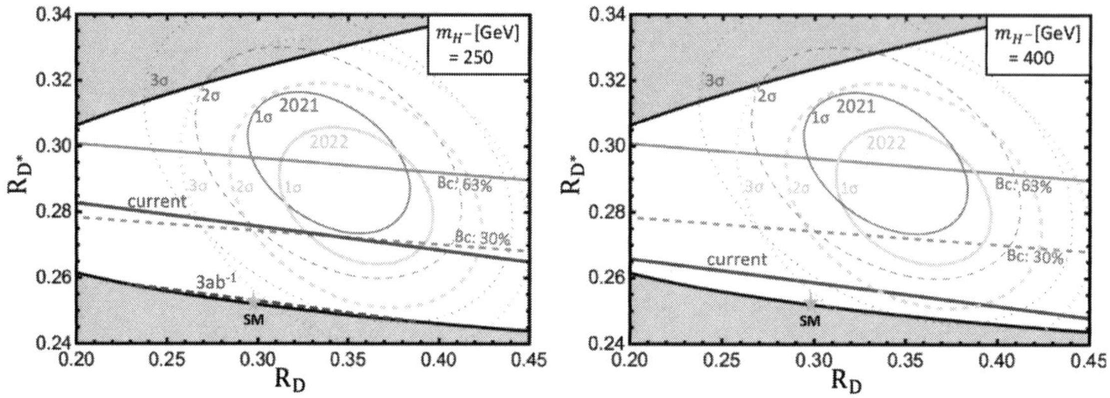

Figure 1. The impact of the $\tau\nu + b$ search on the charged Higgs with $m_{H^-} = 250\,\text{GeV}$ (**left**) and $m_{H^-} = 400\,\text{GeV}$ (**right**). The basic figures are taken from Ref. [17] and the new experimental world average is overlaid with a cyan contour. See the main text for the detail.

$$BR(\bar{B}_s^0 \to D_s^{*+}\pi^-)^{\text{exp}}_{\text{SM}} = \begin{cases} (2.0 \pm 0.5) \times 10^{-3}, \\ (4.46 \pm 0.22) \times 10^{-3}, \end{cases} \quad (6)$$

where $V_{cb} = 0.0397$ is used and transition form factors are taken from Ref. [6]. If we use the inclusive V_{cb} which is larger than than the above V_{cb} by about 5%, the deviations coherently get further larger. The higher order corrections are estimated to be $O(1)$% at most and thus not large enough to fill the gap. The naive size of the discrepancies are 3~5$\,\sigma$. To reproduce the measured branching ratio about 10~20% negative shift at the amplitude level is necessary. Even if we consider the contribution from the final state quasi elastic rescattering, the global fit including color suppressed decays found that similar size of the negative shift is still necessary [11]. It is worth mentioning that the similar tendency is also observed in $\bar{B} \to DV$ modes where V denotes a vector meson,

$$BR(\bar{B}^0 \to D^+K^{*-})^{\text{exp}}_{\text{SM}} = \begin{cases} (4.5 \pm 0.7) \times 10^{-3}, \\ (4.91 \pm 0.41) \times 10^{-3}, \end{cases} \quad (7)$$

$$BR(\bar{B}_s^0 \to D_s^+\rho^-)^{\text{exp}}_{\text{SM}} = \begin{cases} (69 \pm 14) \times 10^{-3}, \\ (105 \pm 7) \times 10^{-3}, \end{cases} \quad (8)$$

however, the large uncertainty do not allow a clear cut conclusion at this stage. Since the mediator needs to couple to light quarks the LHC production cross section could be large.

Another long standing anomaly is known in the muon anomalous magnetic moment (α_μ). The 2021 Fermi-lab result [12] is consistent with the Brookhaven result [13] and deviates from the SM prediction of the white paper (WP) [14] by more than 4σ,

$$\delta\alpha_\mu = \alpha_\mu^{\text{exp}} - \alpha_\mu^{\text{SM}} = (2.51 \pm 0.59) \times 10^{-9}. \quad (9)$$

Although the recent lattice results are more consistent with the experimental result, there is tension with $e^+e^- \to \pi^+\pi^-$ data [15]. If we take this deviation as a hint of the NP, the implied NP scale is also up to $O(1)$ TeV even with the chirality enhancement. Since there is no complete solution to settle the problems and the LHC will accumulate the data regardless of the muon $g-2$ situation, it would be important to propose the independent cross check for the new physics contribution at the high energy colliders.

In the following sections we will introduce the model set up and discuss the LHC sensitivity to probe the interesting scenarios.

2 $R_{D^{(*)}}$ anomaly vs. LHC

The recent LHCb result favors larger (smaller) deviation in R_{D^*} (R_D) compared to the previous world average.[2] This shift has a great impact on the H^- interpretation. Therefore we will update the result in Ref. [16, 17] based on the updated world average.

After integrating out the new particle freedom in general we obtain the following effective Lagrangian

$$\mathcal{L}_{\text{eff}} = -2\sqrt{2}G_F V_{cb}\Big[(1 + C_{V_L})O_{V_L} + C_{V_R}O_{V_R} + C_{S_L}O_{S_L} + C_{S_R}O_{S_R} + C_T O_T\Big], \quad (10)$$

with

$$O_{V_L} = (\bar{c}\gamma^\mu P_L b)(\bar{\tau}\gamma_\mu P_L \nu_\tau), \quad O_{V_R} = (\bar{c}\gamma^\mu P_R b)(\bar{\tau}\gamma_\mu P_L \nu_\tau),$$
$$O_{S_L} = (\bar{c}P_L b)(\bar{\tau}P_L \nu_\tau), \quad O_{S_R} = (\bar{c}P_R b)(\bar{\tau}P_L \nu_\tau),$$
$$O_T = (\bar{c}\sigma^{\mu\nu}P_L b)(\bar{\tau}\sigma_{\mu\nu}P_L \nu_\tau), \quad (11)$$

where projection operators, $P_L = (1-\gamma_5)/2$ and $P_R = (1+\gamma_5)/2$ are defined. The NP contribution is encoded in the Wilson coefficients (WCs) of C_X, normalized by the SM factor of $2\sqrt{2}G_F V_{cb}$. The SM corresponds to $C_X = 0$ for $X = V_{L,R}, S_{L,R}$, and T in this notation.

2.1 Charged scalar

It is well known that the charged Higgs explanation of the discrepancy easily enhances the decay width of the $B_c^- \to \tau\bar{\nu}$ decay. The charged Higgs described bellow can comes from the general two Higgs doublet model (G2HDM) where the both Higgs doublets can couple to

[2] For the comparison see Fig. 1 of Ref. [7].

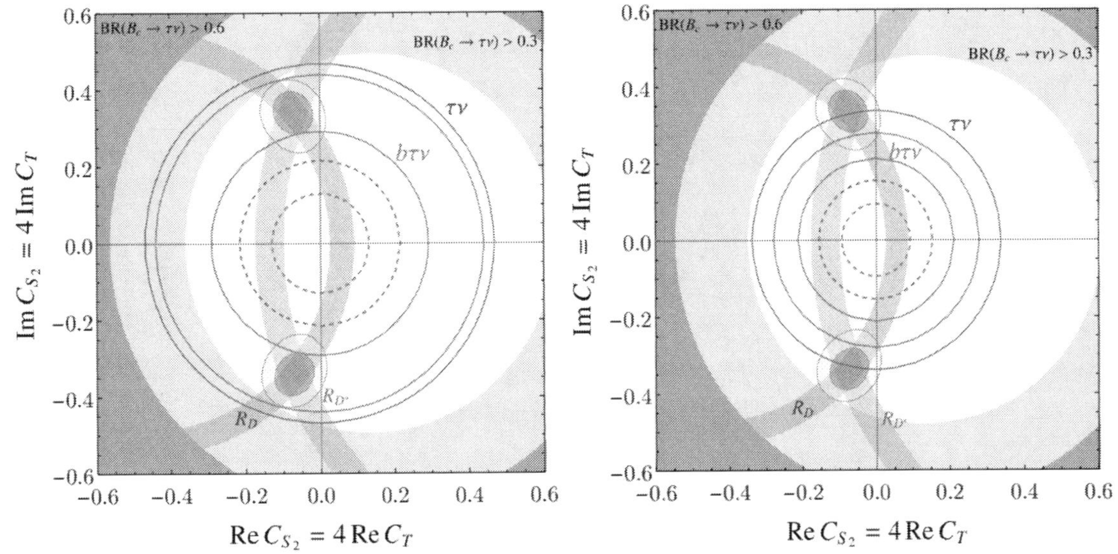

Figure 2. The impact of the $\tau\nu$ and $\tau\nu + b$ searches on the R_2 leptoquark scenario where $m_{R_2} = 2\,\mathrm{TeV}$ (**left**) and $m_{R_2} = 4\,\mathrm{TeV}$ (**right**) are assumed. The figure is taken from Ref. [37]. See the main text for the detail.

all fermions. The upper limits on $BR(B_c^- \to \tau\bar{\nu}) \leq 30\%$ and 10% are derived from the lifetime of the B_c meson [18] and the LEP data [19], respectively. However, it was pointed out that the large charm mass uncertainty and energy dependence of the fragmentation function relax the bound as large as $BR(B_c^- \to \tau\bar{\nu}) \leq 60\%$ [20]. Recently the reevaluation in Ref. [21] found $BR(B_c^- \to \tau\bar{\nu}) \leq 63\%$ based on the B_c lifetime.

Testing the charged Higgs scenario at the LHC is easier than that for the other scenarios since the larger coupling is necessary to enhance R_{D^*} compared to the other scenarios due to small coefficients.[3] In Ref. [22] authors reinterpreted the bound on the heavy resonance reported by the CMS to derive the constraint on the charged Higgs scenario. It was found that the result with $36\,\mathrm{fb}^{-1}$ can already exclude the explanation with $m_{H^-} \geq 400\,\mathrm{GeV}$ where the data is available. In light of the relaxed bound from the B_c decay, Ref. [16] revisited the light mass window $180\,\mathrm{GeV} \leq m_{H^-} \leq 400\,\mathrm{GeV}$ with the $\tau\nu$ constraint based on Run 1 data, the stau constraint and low mass di-jet constraint at the LHC. However, it turned out that this mass region is difficult to fully cover the interesting parameter region even at the high luminosity LHC. More recently it is shown that requiring an additional b-tagged jet in a low mass $\tau\nu$ resonance search, which is not performed experimentally, can enhance the signal to back ground ratio [17]. In a conventional $\tau\nu$ resonance search, the W boson tail consists of a dominant background. By the requiring the additional b-jet, we can suppress the valence quark contribution. For instance $ug \to W^*b \to \tau\nu + b$ contribution is suppressed by the factor of $|V_{ub}|^2$. Furthermore the $j \to b$ miss-tagged contribution in $gu \to W^*j \to \tau\nu + j$ is suppressed by the small miss tagging rate of $O(10^{-3})$. As a result, the huge background reduction of a factor of $O(100)$ is observed while the signal reduction is found to be of a factor.

In Fig. 1, the sensitivity with the current luminosity and the projected sensitivity with the HL-LHC are shown in blue solid and dashed lines on the R_D vs R_{D^*} plane. The area above the line can be probed and the dashed line is omitted on the right since the line almost degenerates with the SM prediction. The charged Higgs mass is assumed to be 250 (400) GeV on the left (right) panel. The world average of the $R_{D^{(*)}}$ data at 1, 2 and 3σ are shown by the solid, dashed and dotted contours. The grey shaded region is out of the prediction within the model. The SM prediction is indicated by a yellow star, and the horizontal magenta solid (dashed) line corresponds to $\mathrm{BR}(B_c \to \tau\nu) = 63\,(30)\%$ as a comparison.

The figure shows that the collider prospect with $139\,\mathrm{fb}^{-1}$ of the data can cover the broader range and judge the $1\,\sigma$ explanation of the anomaly within the model.

2.2 Leptoquarks

There are three types of the leptoquarks, $X = \mathrm{R}_2$, S_1 and U_1 which can explain the discrepancy. The charge assignment under the SM gauge symmetry $(\mathrm{SU}(3)_c, \mathrm{SU}(2)_L)_{\mathrm{U}(1)_Y}$ is given as $(3, 2)_{7/6}$ for R_2, $(\bar{3}, 1)_{1/3}$ for S_1 and $(3, 1)_{3/2}$ for U_1, respectively.

Different from the charged Higgs case, a leptoquark predicts the non resonant $\tau\nu$ signature and the high p_T region is sensitive to the new physics effect since the smaller background contribution is expected there. The significant leptoquark mass dependence is pointed out even with the t-channel mediator because the large momentum exchange, corresponding to the large Mandelstem variable t, is necessary to generate the high p_T leptons [36]. Furthermore it is shown that selecting the negatively charged τ can improve the sensitivity. This is because that the density of the up

[3] See, Eq. (2.4) of Ref.[8] for instance.

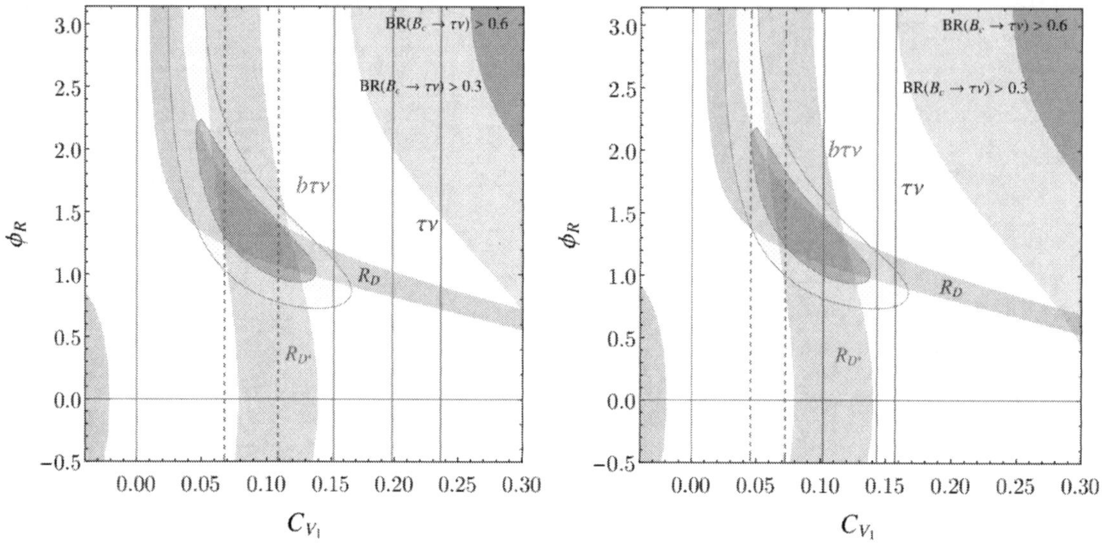

Figure 3. The impact of the $\tau\nu$ and $\tau\nu + b$ searches on the $U(2)$-U$_1$ leptoquark scenario where m_{U_1} = 1.5 TeV (**left**) and m_{U_1} = 4 TeV (**right**) are assumed. The figure is taken from Ref. [37]. See the main text for the detail.

quark is larger than that of down quark in a proton. Also the additional b-tagged helps to enhance the sensitivity to leptoquark scenarios by 30-40% [37]. It is noted that the $\tau\bar{\tau}(+b)$ final state also provides the similar sensitivity[38].

In the following we focus on the R$_2$ and U$_1$ leptoquark scenarios as a demonstration. The R$_2$ scenario predicts $C_{S_2} = +4C_T$ at the leptoquark mass scale where C_{S_2} and C_T are Wilson coefficients of a scalar and tensor operators which describes the $b \to c\tau\bar{\nu}$ transition. In Fig. 2, the constraint and prospect in $\tau\nu$ and $\tau\nu + b$ modes are shown with m_{R_2} = 2 TeV(left) and m_{R_2} = 4 TeV(right). The regions outside the blue and red lines can be probed with the $\tau\nu$ and $\tau\nu + b$ signature, respectively, where the solid and dashed lines correspond to the prospect based on 139 fb^{-1} and 3 ab^{-1} of the data. The magenta line shows the bound with the previous CMS result with 36 fb^{-1} derived in Ref.[36]. It is noted that the CMS found the smaller event number compared to their expectation, and hence it results in the stringent constraint. The darker and lighter gray shaded regions are constrained by BR($B_c \to \tau\nu$) $\leq$ 0.6 and BR($B_c \to \tau\nu$) $\leq$ 0.3. The R_D and R_{D^*} anomalies are explained at 1 σ in the blue and green shaded regions, respectively. The combined fits at 1 and 2 σ are shown in orange and yellow, respectively.

It is found that the LHC sensitivity of the $\tau\nu$ search is marginal with the Run 2 full data to probe the $R_{D^{(*)}}$ explanation depending on the mass, while that of the $\tau^\pm\nu + b$ search is enough to probe the 1σ parameter region in both m_{R_2} = 2 TeV and m_{R_2} = 4 TeV cases. Therefore, it is concluded that requiring an additional b-jet is significant to test a leptoquark scenario in light of the $R_{D^{(*)}}$ anomaly.

Furthermore, the LHC is sensitive to the $U(2)$ flavor symmetry based U$_1$ leptoquark scenario. Recently, the UV completion of the model is widely discussed [23–35]. The $U(2)$ flavor symmetry can naturally suppress the interaction between the LQ and 1st generation fermions and evade the stringent bound from $K_L \to e\bar{\mu}$. The scenario predicts $C_{V_L} = 2e^{i\phi_R}C_{S_R}$ at the LQ mass scale where ϕ_R is free parameter of the model. In fact, unlike the scalar LQ scenarios, flavor constraints can be suppressed or avoided. Therefore, the LHC search is significant to probe the model.

In Fig. 3, the $R_{D^{(*)}}$-favored region is compared with the LHC sensitivities and flavor constraints for $M_{\rm LQ}$ = 1.5 TeV (left) and 4 TeV (right) in the $U(2)$-U$_1$ scenario on the (C_{V_1}, ϕ_R) plane. The region in the right-hand side of the vertical (solid/dashed) lines is probed or constrained by the LHC searches. The orange (yellow) region is favored by the measured $R_{D^{(*)}}$ at the 1 σ (2 σ) level. Note that the best fit is given at $\phi_R \simeq \pm 0.42\pi$, implying $C_{S_1}/C_{V_1} \simeq -0.50 \mp 1.94i$. Similar to the R$_2$ LQ model, imaginary component is favored to be large. From this figure, we found that the $R_{D^{(*)}}$-favored region can be fully (mostly) probed by $\tau^\pm\nu + b$ with 3 ab^{-1} of the data for $M_{\rm LQ}$ > 4 TeV (< 4 TeV).

The study for the S_1 leptoquark scenario can be found in Refs. [36, 37]. For the S_1 LQ and G2HDM, single top + di-τ final state would provide the further important cross check of the scenario.

In section summary, it is clear that if the current size of the discrepancy persists even in the future, the interesting new physics scenarios can be probed at the LHC.

3 $b \to cu\bar{q}$ puzzle vs. LHC

The B meson 2-body hadroic decays are described by the following effective Lagrangian,

$$\mathcal{L} = -\frac{G_F}{\sqrt{2}} \sum_q V_{cb} V_{uq}^* \sum_{i=1,2} C_i^q(\mu) Q_i^q(\mu) , \qquad (12)$$

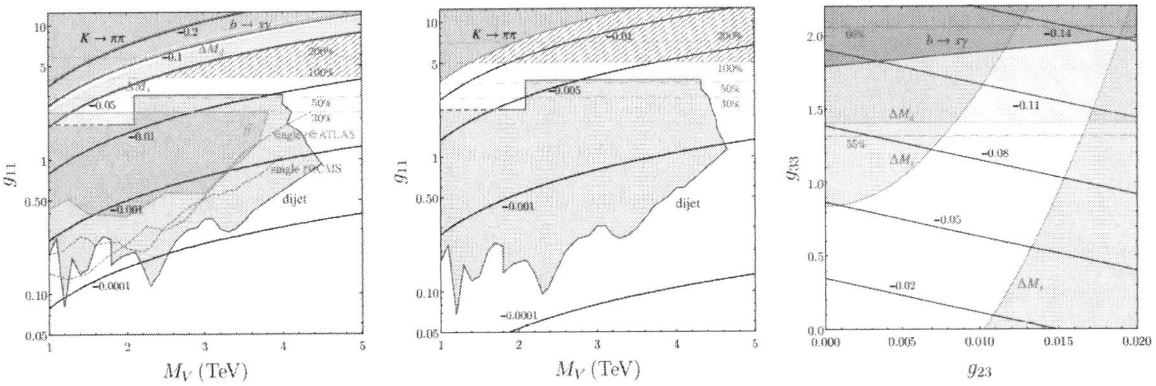

Figure 4. Contours of $C_2^{\rm NP}(m_b)/C_2^{\rm SM}(m_b)$ are presented in black. The figure taken from Ref. [41]. The puzzle can be explained at 2σ level in the yellow bands. The blue and orange shaded regions are excluded by the di-jet and $t\bar{t}$ searches at 95% CL, respectively. The regions above the dashed lines are excluded by the single t searches in the NWA. Furthermore, the gray, red, green, and purple shaded regions are constrained by $K \to \pi\pi$, ΔM_s, ΔM_d, and $b \to s\gamma$, respectively. The dotted line indicates Γ_V/m_V and the red-hatched regions represent $\Gamma_V/m_V > 100\%$. **Left:** scenario $\mathcal{A}$. $g_{33} = -g_{11}$ is taken. **Middle:** scenario BR. $g_{23} = -0.01(M_V/\text{TeV})$ is taken. **Right:** scenario C. $M_V = 1$ TeV and $g_{11} = -3.6$ is fixed which already results in $\Gamma_V/m_V \geq 52\%$.

with the left-handed current-current operators in the CMM basis [39],

$$Q_1^q = (\bar{c}\gamma^\mu T^a P_L b)(\bar{q}\gamma_\mu T^a P_L u)\,, \qquad (13)$$

$$Q_2^q = (\bar{c}\gamma^\mu P_L b)(\bar{q}\gamma_\mu P_L u)\,, \qquad (14)$$

where $q = d, s$. T^a is the $\mathrm{SU}(3)_C$ generator, and V is the Cabibbo-Kobayashi-Maskawa matrix. The suppression of the BR is favored in $B \to DP$, $B \to D^*P$ and $B \to DV$ modes.

Furthermore in a parton level $b \to cu\bar{s}$ and $b \to cu\bar{d}$ require the similar destructive effects. Therefore, if we take those coherent discrepancies as a hint of the NP, it is natural to consider that the new physics has similar flavor and Lorentz structure as the SM. Given that the extension of the gauge sector, adding an additional $SU(2)$ group would be interesting. We consider an extended electroweak gauge group $\mathrm{SU}(2)_1\times\mathrm{SU}(2)_2\times\mathrm{U}(1)_Y$ with heavy vector-like fermions produces heavy gauge bosons, $W'^{\pm}$ and Z', interacting with the left-handed SM fermions with a non-trivial flavor structure [40, 41]. These flavor structures are controlled by the number of generations of the vector-like fermions and mixings between the SM fermions and vector-like fermions. The heavy gauge boson interactions are

$$\mathcal{L} = +\frac{g_{ij}}{2} Z'_\mu \bar{d}^i_L \gamma^\mu d^j_L - \frac{\left(VgV^\dagger\right)_{ij}}{2} Z'_\mu \bar{u}^i_L \gamma^\mu u^j_L - \frac{(Vg)_{ij}}{\sqrt{2}} W'^+_\mu \bar{u}^i_L \gamma^\mu d^j_L + \text{h.c.}\,, \qquad (15)$$

where u_L, d_L are the mass eigenstates, and a coupling g_{ij} is defined in the d_L basis. In the following, we will take $M_{W'} = M_{Z'} = M_V$ for simplicity.

Integrating out W' contribution is obtained as

$$C_2^{q,W'}(M_V) = \frac{1}{4\sqrt{2}G_F M_V^2}\frac{(Vg)_{23}(Vg)^*_{1q}}{V_{cb}V^*_{uq}}\,. \qquad (16)$$

New physics contributions to the Wilson coefficients, $C_1^{q,\rm NP}$ and $C_2^{q,\rm NP}$, become involved at the new physics scale Λ. These values are modified by the renormalization-group (RG) evolution from Λ down to the hadronic scale m_b.

It is found that a universal destructive shift in the SM contributions is favored in the $b \to c\bar{u}q$ puzzle, which corresponds to $C_2^{d,W'} = C_2^{s,W'} = C_2^{W'}$ and

$$\frac{C_2^{W'}(m_b)}{C_2^{\rm SM}(m_b)} \simeq -0.2 \sim -0.1, \qquad (17)$$

where $C_2^{\rm SM}(m_b) \simeq 1$.

In order to generate a desired shift in both $b \to c\bar{u}d$ and $b \to c\bar{u}s$, a SM-like flavor structure in $(Vg)_{1q}$ is required, and hence g_{11} should be non zero. Also another non-zero entry of g_{33} or g_{23} is necessary to produce $C_2^{q,W'}$. Three scenarios are considered in order:

- scenario $\mathcal{A}$, $g_{11} \times g_{33} \neq 0$ and $g_{23} = 0$,
- scenario BR, $g_{11} \times g_{23} \neq 0$ and $g_{33} = 0$,
- scenario C, $g_{11} \times g_{23} \times g_{33} \neq 0$.

The resultant parameter space and the relevant constraints are shown in Fig. 4 for each scenario.

In scenario $\mathcal{A}$, $C_2^{q,W'}$ is proportional to V_{cb} and thus the large couplings and/or lower NP scale are necessary. For instance when $|g_{11}| = |g_{33}| = 2$ is assumed $M_V \simeq 1$ TeV is necessary to fit the data. Even if the flavor violating coupling g_{23} is vanishing, there is W-W' box contribution to ΔM. Also $b \to s\gamma$ and $K \to \pi\pi$ as well as LHC searches can constrain the model parameter space since new gauge bosons considerably couple to valence quarks. More precisely, di-jet, $t\bar{t}$ and single top searches set upper limit on the couplings as long as the particle width does not exceed the experimental assumptions. Currently the maximum width to mass ratio considered in di-jet and $t\bar{t}$ are 55% and 30%. It is noted that the result of single top search is reported based on narrow width approximation

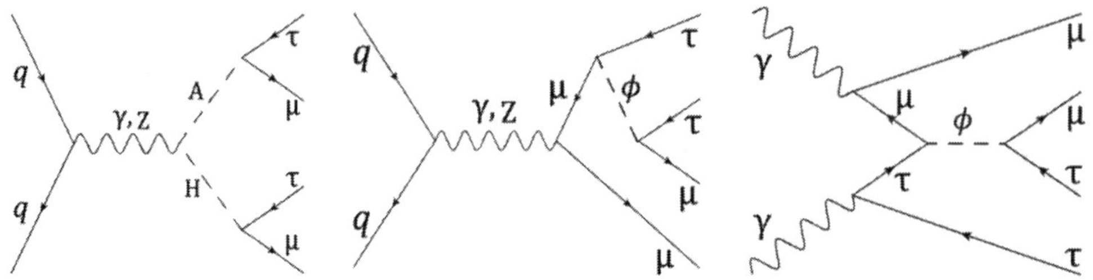

Figure 5. Representative Feynman diagrams that contribute to the $\mu^{\pm}\mu^{\pm}\tau^{\mp}\tau^{\mp}$ signal at the LHC. The left corresponds to electroweak pair production channel. The middle and right diagrams correspond to the single production process where ϕ denotes A or H. In addition, there are also those obtained by exchanging μ and τ which are included in our numerical calculation.

Field	H_1	H_2	(L_e, L_μ, L_τ)	(e_R, μ_R, τ_R)
SM gauge	$(1,2)_{1/2}$	$(1,2)_{1/2}$	$(1,2)_{-1/2}$	$(1,1)_{-1}$
$\mathbb{Z}_4$	1	-1	$(1, i, -i)$	$(1, i, -i)$

Table 1. Relevant field content and charge assignment of the model. The notation of SM gauge quantum numbers is given as $(\mathrm{SU}(3)_c, \mathrm{SU}(2)_L)_{\mathrm{U}(1)_Y}$. H_1 is a SM-like doublet and H_2 is an additional doublet which includes H, A and $H^{\pm}$.

(NWA). Once the particle width gets larger and exceeds the value that experimentally assumed, we can not apply the cross section limit directly and more dedicated analysis is necessary. It is found that if one allows the broad width regime $C_2^{\mathrm{NP}}(m_b)/C_2^{\mathrm{SM}}(m_b) \simeq -0.05$ is possible. Otherwise the allowed shift is less than sub percent.

In scenario *BR*, non zero g_{23} induces ΔM_s at tree level. Also di-jet resonance searches constrains the model. As a result, the possible deviation is $C_2^{\mathrm{NP}}(m_b)/C_2^{\mathrm{SM}}(m_b) \simeq -0.01$ even if one allows the broad width regime.

In scenario *C*, we will demonstrate how large deviation could be possible within the broad width regime, otherwise it is obvious that the possible shift is less than 1%. Interestingly the cancellation between W' box and tree Z' contributions occurs and the stringent constraint from ΔM_s can be relaxed. Consequently $C_2^{\mathrm{NP}}(m_b)/C_2^{\mathrm{SM}}(m_b) \simeq -0.10$ is possible.

It is noted again that the more dedicated collider analysis is necessary to check this parameter region. More specifically, the cross section limit as a function of the minimal m_{jj} would be helpful where m_{jj} is the invariant mass of a pair of the jets. In any case collider constraints are crucial to the new physics interpretation of the discrepancy. Other new physics scenarios are also studied in Ref. [42] and the available mass window in a charged Higgs scenario is pointed out. However, the low mass di-jet + initial state radiation jet search, for instance Ref. [43], would be enough to judge the window mass region.

4 Muon $g-2$ anomary vs. LHC

It is known that the discrepancy is of the same order as the electroweak contribution, i. e. a new $\mathcal{O}(100)$ GeV weakly coupled particle can explain the discrepancy. This fact implies that in order to explain the discrepancy in terms of NP, some enhancement mechanism in the NP contribution to a_μ is necessary to evade the current LHC bound. A popular method to enhance the contribution is the introduction of a new flavor-violating particle. The dipole operator underlying $g-2$ requires a chirality flip, which corresponds to the muon mass within flavor-conserving scenarios. A one-loop contribution involving a $\mu\tau$ lepton flavor-violating (LFV) particle is instead enhanced by a factor of $m_\tau/m_\mu \simeq 17$ [44–47]. This mechanism can lift the mass scale of the new particle by more than a factor of four. However, LFV interactions are stringently constrained and easily spoil the model if the particle also has other couplings. Therefore one needs to ensure the absence of flavor-diagonal couplings for the m_τ enhanced muon $g-2$ solution to be viable. This specific coupling alignment can be realized by a discrete $\mathbb{Z}_4$ flavor symmetry within a 2HDM [48]. See Tab. 1 for the charge assignment. The gauge charge assignments of other SM fields, e. g. quarks, are the same as in the SM, and they trivially transform under $\mathbb{Z}_4$.[4] In this model the $g-2$ contribution is proportional to the $\mu\tau$ LFV Yukawa couplings ($\rho_e^{\mu\tau}$ and $\rho_e^{\tau\mu}$) and the mass difference of the additional neutral scalars which is proportional to the potential coupling λ_5.

In the $\mathbb{Z}_4$ limit the LFV τ decays are not allowed and thus it is not easy to test the scenario at the intensity frontier. Since the new scalars are quark-phobic, their production cross section at the LHC is not large. However, the unique coupling structure predicts that the neutral scalars decay into $\mu^{\pm}\tau^{\mp}$. Previously we pointed out the smoking-gun signature of a $\mu^{\pm}\mu^{\pm}\tau^{\mp}\tau^{\mp}$ final state via electroweak scalar pair production (left of Fig. 5) with a special focus on the case where all Yukawa and scalar potential couplings are smaller than one [45]. We argued that the full Run 2 data set can test the model up to 500 GeV scalar mass thanks to the very unique double $\mu\tau$ LFV resonance nature of the signal events. However, if we accept relatively large coupling of $\mathcal{O}(1)$, the model can still explain the discrepancy with 1 TeV scalars.

Therefore we extend previous study with the perturbative unitarity constraint which gives an upper limit on the additional scalar mass as

$$m_A \lesssim 1650\,\mathrm{GeV}, \tag{18}$$

[4]In order to obtain realistic neutrino masses and mixings the model needs to be extended.

 https://doi.org/10.1051/epjconf/202328901007

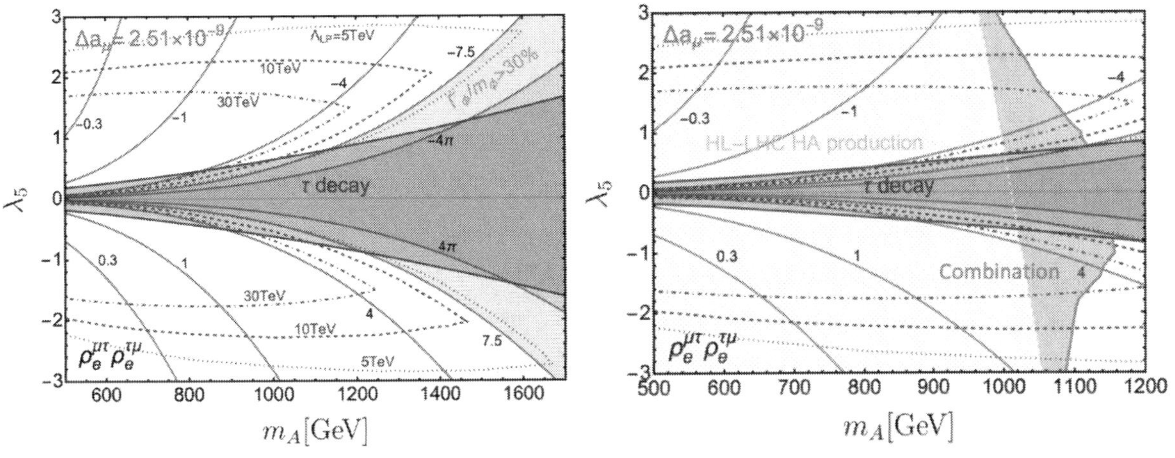

Figure 6. The black contours show the value of $\rho_e^{\mu\tau}\rho_e^{\tau\mu}$ to explain the central value of a_μ in the m_A vs. λ_5 plane. The figure is taken from Ref. [51]. The blue region is excluded by the lepton flavor universality of τ decays, BR($\tau \to \mu\nu\bar{\nu}$)/BR($\tau \to e\nu\bar{\nu}$). The orange region corresponds to $\Gamma_\phi/m_\phi \geq 30\%$. The purple contours depict the cutoff scale of the model. On the right panel the collider sensitivities are overlaid. The cyan region can be tested with 3 ab^{-1} of the electroweak pair production, while the magenta region can additionally be probed by including the single production channel. The blue and orange regions and purple contours are the same as in the left. $\Delta a_\mu = 2.51 \times 10^{-9}$ is fixed and $|\rho_e^{\mu\tau}| = |\rho_e^{\tau\mu}|$ is assumed in the figure.

and evaluate the search potential [51].[5] Fig. 6 (left) summarize the available parameter space.

We use MadGraph5_aMC@NLO [49] with NNPDF3.1luxQED [50] to calculate the signal cross sections. To account for the minimal kinematic cuts, $|p_T^\mu|, |p_T^\tau| \geq 20$ GeV, $|\eta_\mu|, |\eta_\tau| \leq 2.7$, and $\Delta R \geq 0.1$ are imposed. We assume focus on the hadronic decaying τ. Our signal of same sign $\mu^\pm\mu^\pm$ and $\tau^\mp\tau^\mp$ pairs of which $\mu^\pm\tau^\mp$ forms a resonance is very distinctive. Specifically, due to the resonance structure, the final-state leptons are very energetic. Therefore we can safely assume the SM background (SMBG) to be negligible. Based on the Poisson statistics, we evaluated the sensitivity at 95% confidence level.

The cyan region in Fig. 6 show the HL-LHC sensitivity based on the EW pair production. The sensitivity is asymmetric in λ_5, since $\lambda_5 \geq 0$ corresponds to $m_H \geq m_A$ and thus the production cross section will be smaller compared to $m_H \leq m_A$. The HL-LHC data of 3 ab^{-1} would be sensitive to mass scales of $m_A \simeq 960$ GeV. It is worth mentioning that the pair production channel is powerful since once the neutral scalars are produced they dominantly decay into $\mu\tau$, as long as a sizable τ-mass enhanced contribution to a_μ is postulated. Nevertheless there is a mass gap between the sensitivity and the theoretical upper limit of Eq. (18). The loss of sensitivity for larger m_A mainly comes from two factors: the contributing coupling constant is a weak gauge coupling which is independent of Δa_μ and the production cross section is suppressed by the heaviness of the pair-produced scalars.

One possible way to extend the LHC reach to our model is to include the single heavy scalar production channels corresponding to the middle and right diagrams of Fig. 5. Especially for $\mathcal{O}(1)$ TeV lepto-philic particles the inclusion of the photon initiated process is important [46]. There are two terms in the single scalar production amplitude to generate $\mu^\pm\mu^\pm\tau^\mp\tau^\mp$ events. One is proportional to $(|\rho_e^{\mu\tau}|^2 + |\rho_e^{\tau\mu}|^2)$ which vanishes in the $m_A = m_H$ limit. On the other hand the term proportional to $\rho_e^{\mu\tau}\rho_e^{\tau\mu}$ does not disappear in this limit. Since the heavy mass scenario requires a large product of the LFV Yukawa couplings, the second contribution will be important for $\mathcal{O}(1)$ TeV scalars.

In Fig. 6 (right) the magenta region shows the combined sensitivity. The magenta regions can additionally be covered by including also the single scalar production. We note that there is also a non-resonant signal contribution which comes from t-channel A/H exchange. Since the lepton p_T in this case is generally small and we are interested in the high p_T region where the BG is negligible, this contribution is separated and subtracted to evaluate the sensitivity. We find that the inclusion of the single production process can improve the experimental reach by 130 and 60 GeV for $|\lambda_5| \simeq 1$ and 2, respectively, when the Yukawa couplings are large. This still leaves a gap between the experimental reach and theoretical upper limit.

In order to further boost the sensitivity to the large-mass scenario it is important to increase the center of mass energy from $\sqrt{s} = 14$ TeV to, for instance, 27 TeV [52]. Again the reach of the $\mu^\pm\mu^\pm\tau^\mp\tau^\mp$ channel can be extended by including the single production process. As a result all the available parameter region can be probed with the high energy LHC with 3 ab^{-1} of the data.

It is noted that adding $\mathbb{Z}_4$ charged singlet scalar could solve the dark matter problem [53], $\mu^\pm\mu^\pm\tau^\mp\tau^\mp$ channel can also cover the interesting parameter space.

[5] We required that the theory remains perturbative up to at least 5 TeV. If we require the theory to be perturbative up to 30 TeV, the upper limit is given as $m_A \leq 1250$ GeV

5 Conclusion

After the SM-like Higgs boson discovery there is no established NP signal at the LHC. On the other hand there are several statistically large discrepancies between the SM prediction and measurement known in the flavor physics. If we take those deviations as a hint of the NP, $O(1)$ TeV new particles are implied. It is expected that the LHC plays a very important role in independently confirming/excluding the those scenarios. To improve the search potential and probe interesting scenarios, additional b-jet tagging technique, mediator mass dependence in t-channel processes and distinctive $\mu^\pm\mu^\pm\tau^\mp\tau^\mp$ final state are discussed. We found that the future LHC have great potential and allow us to judge new physic interpretations.

Acknowledgements

I would like to thank the organizers of the Workshop "Flavour changing and conserving processes 2022" for the kind invitation to participate to this fifth edition of FCCP Workshop, and for choosing the perfect place for such a great meeting. This work is supported by the Deutsche Forschungsgemeinschaft (DFG, German Research Foundation) under grant 396021762-TRR 257.

References

[1] A. Crivellin and M. Hoferichter, Science **374**, no.6571, 1051 (2021)
[2] M. Kohda, T. Modak and A. Soffer, Phys. Rev. D **97**, no.11, 115019 (2018)
[3] B. C. Allanach, J. M. Butterworth and T. Corbett, Eur. Phys. J. C **81**, no.12, 1126 (2021)
[4] A. Azatov *et al.* JHEP **10**, 149 (2022)
[5] Y. Amhis *et al.* [HFLAV], [arXiv:2206.07501 [hep-ex]].
[6] S. Iguro and R. Watanabe, JHEP **08**, no.08, 006 (2020)
[7] S. Iguro, T. Kitahara and R. Watanabe, [arXiv:2210.10751 [hep-ph]].
[8] S. Iguro *et al.* JHEP **02**, 194 (2019)
[9] M. Bordone, N. Gubernari, T. Huber, M. Jung and D. van Dyk, Eur. Phys. J. C **80**, no.10, 951 (2020)
[10] R. L. Workman *et al.* [Particle Data Group], PTEP **2022**, 083C01 (2022)
[11] M. Endo, S. Iguro and S. Mishima, JHEP **01**, 147 (2022)
[12] B. Abi *et al.* [Muon g-2], Phys. Rev. Lett. **126**, no.14, 141801 (2021)
[13] G. W. Bennett *et al.* [Muon g-2], Phys. Rev. D **73**, 072003 (2006)
[14] T. Aoyama *et al.* Phys. Rept. **887**, 1-166 (2020)
[15] M. Knecht, these proceedings. See also other proceedings.
[16] S. Iguro, Phys. Rev. D **105**, no.9, 095011 (2022)
[17] M. Blanke, S. Iguro and H. Zhang, JHEP **06**, 043 (2022)
[18] R. Alonso, B. Grinstein and J. Martin Camalich, Phys. Rev. Lett. **118**, no.8, 081802 (2017)
[19] A. G. Akeroyd and C. H. Chen, Phys. Rev. D **96**, no.7, 075011 (2017)
[20] M. Blanke, *et al.* Phys. Rev. D **99**, no.7, 075006 (2019)
[21] J. Aebischer and B. Grinstein, JHEP **07**, 130 (2021)
[22] S. Iguro, Y. Omura and M. Takeuchi, Phys. Rev. D **99**, no.7, 075013 (2019)
[23] L. Di Luzio, A. Greljo and M. Nardecchia, Phys. Rev. D **96**, no.11, 115011 (2017)
[24] M. Bordone *et al.* Phys. Lett. B **779**, 317-323 (2018)
[25] M. Blanke and A. Crivellin, Phys. Rev. Lett. **121**, no.1, 011801 (2018)
[26] A. Greljo and B. A. Stefanek, Phys. Lett. B **782**, 131-138 (2018)
[27] L. Di Luzio *et al.* JHEP **11**, 081 (2018)
[28] C. Cornella, J. Fuentes-Martin and G. Isidori, JHEP **07**, 168 (2019)
[29] S. Balaji, R. Foot and M. A. Schmidt, Phys. Rev. D **99**, no.1, 015029 (2019)
[30] S. Balaji and M. A. Schmidt, Phys. Rev. D **101**, no.1, 015026 (2020)
[31] J. Fuentes-Martín and P. Stangl, Phys. Lett. B **811**, 135953 (2020)
[32] S. Iguro *et al.* Phys. Rev. D **104**, no.7, 075008 (2021)
[33] S. F. King, JHEP **11**, 161 (2021)
[34] S. Iguro *et al.* JHEP **07**, 022 (2022)
[35] M. Fernández Navarro and S. F. King, [arXiv:2209.00276 [hep-ph]].
[36] S. Iguro, M. Takeuchi and R. Watanabe, Eur. Phys. J. C **81**, no.5, 406 (2021)
[37] M. Endo *et al.* JHEP **02**, 106 (2022)
[38] [CMS], CMS-PAS-HIG-21-001.
[39] K. G. Chetyrkin, M. Misiak and M. Munz, Phys. Lett. B **400**, 206-219 (1997) [erratum: Phys. Lett. B **425**, 414 (1998)]
[40] S. M. Boucenna *et al.* JHEP **12**, 059 (2016)
[41] S. Iguro and T. Kitahara, Phys. Rev. D **102**, no.7, 071701 (2020)
[42]
[42] M. Bordone, A. Greljo and D. Marzocca, JHEP **08**, 036 (2021)
[43] A. M. Sirunyan *et al.* [CMS], JHEP **01**, 097 (2018)
[44] S. Iguro and Y. Omura, JHEP **05**, 173 (2018)
[45] S. Iguro, Y. Omura and M. Takeuchi, JHEP **11**, 130 (2019)
[46] S. Iguro, K. A. Mohan and C. P. Yuan, Phys. Rev. D **101**, no.7, 075011 (2020)
[47] S. Iguro, Y. Omura and M. Takeuchi, JHEP **09**, 144 (2020)
[48] Y. Abe, T. Toma and K. Tsumura, JHEP **06**, 142 (2019)
[49] J. Alwall *et al.* JHEP **07**, 079 (2014)
[50] V. Bertone *et al.* [NNPDF], SciPost Phys. **5**, no.1, 008 (2018)
[51] M. Blanke and S. Iguro, [arXiv:2210.13508 [hep-ph]].
[52] A. Abada *et al.* [FCC], Eur. Phys. J. ST **228**, no.5, 1109-1382 (2019)
[53] K. Asai *et al.* Phys. Rev. D **106**, no.3, 035017 (2022)

Preparations for a search of the muon EDM at PSI

P. Schmidt-Wellenburg[1,*], C. Calzolaio[1], A. Doinaki[1], C. Dutsov[1], M. Giovannozzi[2], T. Hume[1], and F. Trillaud[3] on behalf of the muEDM collaboration.

[1] *Paul Scherrer Institute, 5232 Villigen, Switzerland*
[2] *CERN Beams Department, Esplanade des Particules 1, 1211 Meyrin, Switzerland*
[3] *Instituto de Ingeniería, Universidad Nacional Autónoma de Mexico, 04510 Ciudad de Mexico, Mexico*

Abstract. Electric dipole moments (EDMs) of fundamental particles violate time invariance and the combined symmetry of charge and parity (CP). The existence of a large muon EDM (muEDM) is made plausible by tensions with Standard Model predictions for semileptonic decays of heavy meson measured at LHCb, BaBar, and Belle, as well as the muon's anomalous magnetic moment (AMM). A discovery of the muEDM would manifest CP and lepton flavor universality (LFU) violation, revealing physics beyond the SM (BSM). The most sensitive muEDM search to date provides an upper limit of $1.8 \times 10^{-19}\, e{\cdot}\mathrm{cm}$ (CL 95%), extracted from high-precision data collected to measure the muon AMM. At the Paul Scherrer Institute, we are setting up a dedicated search for the muEDM using, for the first time, the frozen-spin technique to target an ultimate sensitivity better than $6 \times 10^{-23}\, e{\cdot}\mathrm{cm}$. This novel technique increases the sensitivity to EDM-induced spin precession by cancelling the AMM-induced precession with the application of a precisely tuned electric field perpendicular to the muon momentum and the magnetic field. In this configuration, the dominant source of precession is the EDM coupling to the large relativistic electric field in the muon rest frame, generated by its motion in a strong 3 T uniform magnetic field. In a precursor experiment, we will apply the frozen-spin technique in a compact solenoid demonstrating a sensitivity of better than $3 \times 10^{-21}\, e{\cdot}\mathrm{cm}$, probing uncharted and otherwise inaccessible territory in BSM theories.

1 Introduction

Electric dipole moments (EDMs) of fundamental particles violate time-reversal and, invoking the CPT-theorem [1], the combined symmetry of charge conjugation and parity inversion (CP). More than 70 years ago, E.M. Purcell, N.F. Ramsey and their student J.H. Smith [2] searched for an EDM of the neutron for the first time. The second search was performed using muons, measuring their longitudinal decay asymmetry after being deflected by a dipole magnet [3] resulting in an upper limit of $2.8 \times 10^{-15}\, e{\cdot}\mathrm{cm}$ (CL 95%). Since then, many searches around the world with increasing sensitivity on neutrons, atoms, and molecules [4, 5] have been concluded so far without detection.

The latest search for a muon EDM using Brookhaven National Lab's g-2 storage ring data [6], and searching for a vertical oscillation, longitudinal with respect to the main magnetic field, resulted in the current upper limit of $1.8 \times 10^{-19}\, e{\cdot}\mathrm{cm}$ (CL 95%) [7]. Using existing data from the measurements of the anomalous magnetic moment (AMM), (g-2)/2, of the muon to search for the muon EDM is economical but cannot overcome the intrinsic limitation due to the measurement of an oscillation amplitude instead of its frequency or phase. Within the historically predominant paradigm, in which lepton flavor universality and minimal flavor violation were assumed also almost universally in BSM physics, a dedicated muon EDM search did not seem attractive. It was assumed that the stringent limit on the electron EDM deduced from measurements using atoms or molecules, e.g. thorium oxide molecules $d_e < 1.1 \times 10^{-29}\, e{\cdot}\mathrm{cm}$ (CL 90%) [8], could be naively rescaled by the ratio m_μ/m_e, resulting in a limit of $d_{\mu\leftarrow e} < 2.3{\times}10^{-27}\, e{\cdot}\mathrm{cm}$ (CL 90%), which is eight orders of magnitude better than the direct limit $d_\mu < 1.5 \times 10^{-19}\, e{\cdot}\mathrm{cm}$ (CL 90%). Since no new particles elucidating BSM physics have been found at the LHC yet [9, 10], the assumptions of this bygone paradigm are being relinquished. This motivates the study of flavor-violating processes and hence a direct search for the muon EDM, independent of the electron EDM limit.

When indirect effects are tested [11], high-precision measurements have access to energies that exceed the reach of the LHC. Intriguingly, the most substantial evidence for BSM appears in semileptonic B-meson decays with heavy leptons, especially with muons [12–14], at LHCb, Belle, and BaBar, violating lepton flavor universality (LFU), and in the evidence of the muon g-2 discrepancy with SM expectations [15, 16]. These striking hints for new physics are incompatible with minimal flavor violation (MFV) in the lepton sector [17]. In effective-field theories, the imaginary part of the Wilson coefficient, of

* e-mail: μphilipp.schmidt-wellenburg@psi.ch

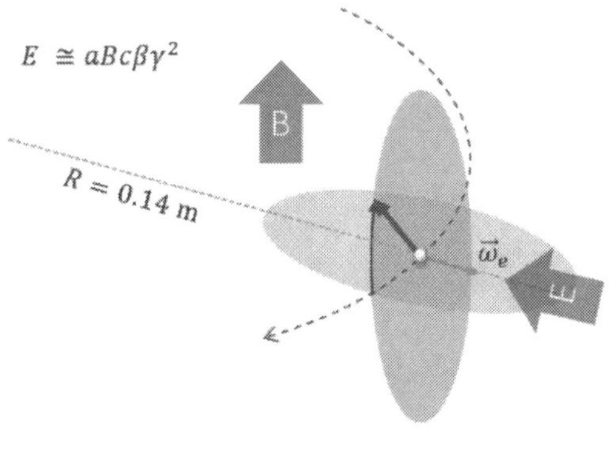

Figure 1. Illustration of the spin precession due to the AMM and the EDM of the muon.

which the real part gives rise to the g-2 of a lepton, also gives rise to the EDM [18, 19]. This intrinsic connection of the EDM to the g-2 and the tantalising evidence for LFU violation can only be tested by a dedicated high-sensitivity search for the muon EDM.

2 The frozen-spin technique in a compact solenoid

The EDM of a charged particle with charge q and spin $\hbar/2$, $\vec{s}$, can be described as $\vec{d} = \eta q\vec{s}/(2mc)$ and leads to a two-fold energy splitting when exposed to an electric field with

$$\Delta E = 2\vec{d} \cdot \vec{E}, \tag{1}$$

corresponding to a transition frequency between the upper and lower energy states of $\omega_e = 2d|E|/\hbar$. As the spin is an axial vector while the electric field is a vector, the scalar product of the two violates time and parity symmetry, and by the CPT theorem [1] also the combined symmetry of charge conjugation and parity (CP).

The spin dynamics of a charged particle with spin $\hbar/2$ in a magnetic and electric field is described by the Thomas-Bargman, Michel, and Telegdi [20, 21] equation,

$$\begin{aligned}\vec{\omega}_a &= \vec{\omega}_c - \vec{\omega}_0 \\ &= -\frac{q}{m}\left[a\vec{B} - \frac{a\gamma}{(\gamma+1)}\left(\vec{\beta}\cdot\vec{B}\right)\vec{\beta} - \left(a + \frac{1}{1-\gamma^2}\right)\frac{\vec{\beta}\times\vec{E}}{c}\right],\end{aligned} \tag{2}$$

where $\vec{\omega}_c$ and $\vec{\omega}_0$ are the relativistic cyclotron and Larmor frequency, respectively and $a = (g-2)/2$ is the anomalous magnetic moment. This will result in a precession of the spin aligned with the magnetic field, $\vec{\omega} \parallel \vec{B}$, indicated by the gray ellipse in Fig. 1.

An EDM $\vec{d}$ will result in an additional oscillation,

$$\vec{\omega}_e = \frac{\eta q}{2m}\left[\vec{\beta}\times\vec{B} + \frac{\vec{E}}{c} - \frac{\gamma c}{(\gamma+1)}\left(\vec{\beta}\cdot\vec{E}\right)\vec{\beta}\right], \tag{3}$$

perpendicular to momentum and magnetic field. To employ the frozen-spin technique, the electric field must be tuned so that

	phase I	phase II
Muon flux (μ^+/s)	4×10^6	1.2×10^8
Channel transmission	0.03	0.005
Injection efficiency	0.017	0.60
Muon storage rate (1/s)	2×10^3	360×10^3
Gamma factor γ	1.04	1.56
Magnetic field (T)	3	3
Electric field (kV/cm)	3	20
e^+ detection rate (1/s)	500	90×10^3
Detections per 200 days	8.64×10^9	1.5×10^{12}
Mean decay asymmetry	0.3	0.3
Initial polarization P_0	0.95	0.95
Sensitivity (e·cm)	$< 3 \times 10^{-21}$	$< 6 \times 10^{-23}$

Table 1. Annual statistical sensitivity of the muon EDM measurement of phase I and II.

$$a\vec{B} = \left(a - \frac{1}{\gamma^2 - 1}\right)\frac{\vec{\beta}\times\vec{E}}{c}, \tag{4}$$

with $|E| \approx a|B|c\beta\gamma^2$, which, in the nominal case of $\vec{B} \perp \beta$, results in $\vec{\omega}_a = 0$ and the spin is permanently aligned, i.e. frozen, to the momentum. Therefore, in the presence of an EDM one will observe a spin precession $\vec{\omega}_e \parallel \left(\vec{\beta}\times\vec{B}\right)$.

The frozen-spin technique benefits from a continuous build-up of the phase $\omega_e t$, and results in high statistical sensitivities. Two stages are considered at PSI. In the first phase (phase I) we will use surface muons with $p \approx 30\,\mathrm{MeV/c}$, while in the final setup (phase II) we would like to profit from the high flux available with $p = 125\,\mathrm{MeV/c}$ muons from pions decaying in flight. In a magnetic field of 3 T this results in an EDM sensitivity for a single muon of $\sigma(d_\mu) \approx 2 \times 10^{-17}\,e\cdot$cm and $\sigma(d_\mu) \approx 5 \times 10^{-17}\,e\cdot$cm, for a muon beam from decays of pions at rest and in flight, respectively. This, in turn, results in an electric field for the frozen-spin condition of $E_f = 0.3\,\mathrm{MV/m}$ or $E_f = 1.9\,\mathrm{MV/m}$.

In both phases we will inject the muons off-axis along the solenoid field direction. For a high injection rate, the muons need to pass through a dedicated collimation channel, which also includes a magnetic shield made of superconductors, as the acceptance phase space would be extremely small without the magnetic shield as a result of the magnetic mirror effect. The muons then spiral along the magnetic field to the center of the solenoid, where a quadrupole magnetic kick within the weakly focusing field region will twist the muon momentum onto a stable storage orbit. This longitudinal injection, also known as vertical injection, was first described in [22]. It profits from the advantage that the injection channel is far from the magnetically sensitive region of the storage zone and from a long delay, of the order of about 100 ns, between detecting a suitable muon using an entrance detector and the application of the magnetic kick. Table 1 summarizes all the

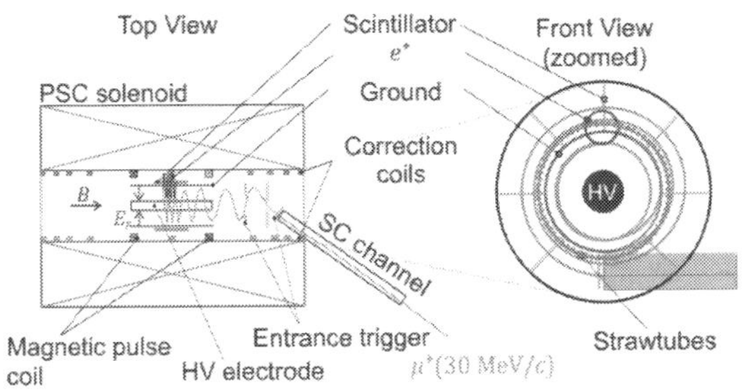

Figure 2. Side view (left) and front view (right) of the demonstrator experiment using an existing solenoid magnet with a field strength of 3 T.

factors necessary for the statistical sensitivity analysis in both phases.

3 Demonstration experiment using an existing solenoid

In the first phase, we intend to use an existing solenoid with a field of 3 T to demonstrate all the techniques necessary for the muon EDM search deploying the frozen-spin technique. A conceptual sketch is shown in Fig. 2. Muons with momentum of $P \approx 30\,\mathrm{MeV/c}$ from the decay of pions at rest with a polarization better than 95 % will be injected into a collimation tube of inner diameter ID= 15 mm and length $\ell \approx 800$ mm within a magnetic shield. While the collimation is the first step in reducing the phase space of the injected beam so as to match it to the tiny acceptance phase space, the magnetic superconducting shield (SC-channel) is essential to transport the muons from the low-field region at the exit of the beamline to the high-field region inside the solenoid. A set of correction coils will reduce the field gradient between the injection area at the exit of the collimation tube to increase the acceptance phase space. Once a muon is injected into the solenoid, a thin entrance scintillator in anticoincidence with a set of active apertures generates an entrance signal for muons within the acceptance phase space, which triggers a magnetic pulse in the center of the solenoid. The 100 ns short quadrupole magnetic pulse (pulse coil) twists the remaining longitudinal momentum, along the solenoid field, of the incident muon in the transverse direction. The muon is hence stored on a stable orbit inside the weakly-focusing field. An electric field $E_f = 3\,\mathrm{kV/cm}$ at the storage orbit is applied between two coaxial electrodes and establishes the frozen-spin technique. A combination of straw tubes and scintillating fiber mats track the decay positron to measure the $g-2$ frequency and the longitudinal asymmetry. The measurement of the $g-2$ frequency, ω_a, will serve as a sensitive magnetic field probe. In addition, we will tune the E field for the frozen-spin condition by measuring ω_a as a function of the applied electric field and interpolate to $\omega_a(E) = 0$.

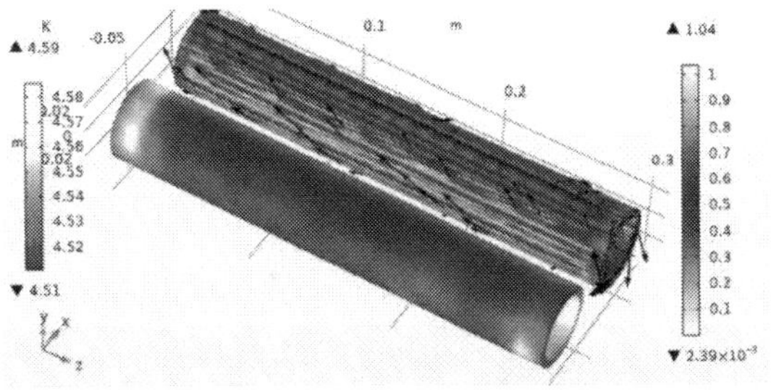

Figure 3. Finite element calculation of an injection tube made of a superconducting sheet.

4 Injection into the solenoid field

In a first step, we will demonstrate the off-axis injection into the solenoid and through a collimation channel inside a magnetic shield. In the fringe field of the solenoid magnet, below 1 T, we will use a thick iron tube as a magnetic shield. Above 1 T and up to 3 T a shield made of superconductor (SC) will be deployed. More than fifty years ago, Firth et al. demonstrated this for a 1.75 T bubble chamber at CERN [23] and it is used today, for example, in the BNL/FNAL (g-2) experiment [24]. The principal idea is that once the superconducting shield is cooled below the critical temperature T_c, ramping of the outside field will induce persistent currents inside the superconductor that counteract with the outside field. Therefore, the field inside the SC shield will remain as low as before the magnet was ramped. This effect persists if the shield is sufficiently thick and the mean lifetime of the shielding current is long enough. Once the field starts to penetrate, the outside field needs to be ramped down, and the superconducting shield can be reset by warming up above the critical temperature. We will test Nb-Ti/Nb/Cu sheets from the Wigner Research Center in Budapest, wound and clamped around a copper tube of ID=15 mm, and a design based on high-temperature superconducting ribbons / tapes wound helically onto a copper tube of the same diameter. Similar tests [25, 26] showed promising results for the Nb-Ti/Nb/Cu sheets, while the HTS ribbon design did not adequately shield against the outside field. We will investigate whether different mounting techniques and more layers of the helical wound HTS result in a sufficient shielding factor. This would be favorable as only liquid nitrogen temperature will be required if the final experiment's magnet also deploys HTS coils. We are currently preparing finite element models of superconducting shields (for illustration, see Fig.,3), and benchmark single layer models against physical prototypes (see Fig. 4). For the HTS shields, we are setting up a test facility using a liquid nitrogen bath and a Helmholtz coil. We will measure the shielding factor by taking the ratio of magnetic fields inside the tubes when the coil is ramped to the maximum field at room temperature and 70 K.

Figure 4. First prototype magnetic shields for the injection channel. On the left, two versions made of high temperature superconducting tape would onto a copper tube. On the right, two tube of standard iron tubes which might be used in the region with fields below 1 T.

5 Conclusion

At PSI we are setting up an experiment to search for the electric dipole moment of the muon using the frozen-spin technique. In the next few years, we will demonstrate essential experimental techniques using an existing solenoid magnet with a field strength of 3 T. For the injection of muons into the high-field region, we will deploy a SC shield made of either Nb-Ti/Nb/Cu sheets or HTS ribbon wound around a collimation tube. In this initial phase, we plan to demonstrate the frozen-spin technique by searching for muon EDM with a sensitivity of better than 3×10^{-21} $e{\cdot}$cm using surface muons. In phase II, using an optimized highly uniform magnetic field and muons with a momentum of at least 125 MeV/c, we aim for a sensitivity of at least 6×10^{-23} $e{\cdot}$cm.

6 Acknowledgments

The authors and the collaboration thank the excellent technical support of M. Meier from the UCN group and R. Deckardt and M. Duda from GFA at PSI. This work is partially financed by the Swiss National Science Fund's grant No.204118 and received funding from the Swiss State Secretariat for Education, Research and Innovation (SERI).

References

[1] G. Lüders, Dan. Mat. Fys. Medd. **28**, 1 (1954)
[2] J.H. Smith, E.M. Purcell, N.F. Ramsey, Phys. Rev. **108**, 120 (1957)
[3] D. Berley, R.L. Garwin, G. Gidal, L.M. Lederman, Phys. Rev. Lett. **1**, 144 (1958)
[4] K. Jungmann, Ann. Phys. **525**, 550 (2013)
[5] T.E. Chupp, P. Fierlinger, M.J. Ramsey-Musolf, J.T. Singh, Rev. Mod. Phys. **91**, 015001 (2019)
[6] G.W. Bennett, B. Bousquet, H.N. Brown, G. Bunce, R.M. Carey, P. Cushman, G.T. Danby, P.T. Debevec, M. Deile, H. Deng et al. (Muon g-2), Phys. Rev. D **73**, 072003 (2006), hep-ex/0602035
[7] G.W. Bennett, B. Bousquet, H.N. Brown, G. Bunce, R.M. Carey, P. Cushman, G.T. Danby, P.T. Debevec, M. Deile, H. Deng et al. (Muon (g-2)), Phys. Rev. D **80**, 052008 (2009), 0811.1207
[8] V. Andreev, D. Ang, D. DeMille, J. Doyle, G. Gabrielse, J. Haefner, N. Hutzler, Z. Lasner, C. Meisenhelder, B. O'Leary et al., Nature **562**, 355 (2018)
[9] J.N. Butler, arXiv e-prints arXiv:1709.03006 (2017), 1709.03006
[10] L. Masetti, A. Collaboration, Nuclear and Particle Physics Proceedings **303-305**, 43 (2018)
[11] M. Raidal, A. Schaaf, I. Bigi, M. Mangano, Y. Semertzidis, S. Abel, S. Albino, S. Antusch, E. Arganda, B. Bajc et al., Eur. Phys. J. C **57**, 13 (2008), 0801.1826
[12] B. Capdevila, A. Crivellin, S. Descotes-Genon, J. Matias, J. Virto, JHEP **2018**, 93 (2018), 1704.05340
[13] W. Altmannshofer, P. Stangl, Eur. Phys. J. C **81**, 952 (2021), 2103.13370
[14] LHCb Collaboration, R. Aaij, C. Abellán Beteta, T. Ackernley, B. Adeva, M. Adinolfi, H. Afsharnia, C.A. Aidala, S. Aiola, Z. Ajaltouni et al., Nature Physics **18**, 277 (2022), 2103.11769
[15] T. Aoyama, N. Asmussen, M. Benayoun, J. Bijnens, T. Blum, M. Bruno, I. Caprini, C.M. Carloni Calame, M. Cè, G. Colangelo et al., Phys. Rep. **887**, 1 (2020), 2006.04822
[16] B. Abi, T. Albahri, S. Al-Kilani, D. Allspach, L.P. Alonzi, A. Anastasi, A. Anisenkov, F. Azfar, K. Badgley, S. Baeßler et al. (Muon $g-2$ Collaboration), Phys. Rev. Lett. **126**, 141801 (2021)
[17] V. Cirigliano, B. Grinstein, G. Isidori, M.B. Wise, Nuclear Physics B **728**, 121 (2005), hep-ph/0507001
[18] J.L. Feng, K.T. Matchev, Y. Shadmi, Nucl. Phys. **B613**, 366 (2001), hep-ph/0107182
[19] A. Crivellin, M. Hoferichter, P. Schmidt-Wellenburg, Phys. Rev. D **98**, 113002 (2018), 1807.11484
[20] L. Thomas, The London, Edinburgh, and Dublin Philosophical Magazine and Journal of Science **3**, 1 (1927)
[21] V. Bargmann, L. Michel, V.L. Telegdi, Phys. Rev. Lett. **2**, 435 (1959)
[22] H. Iinuma, H. Nakayama, K. Oide, K.i. Sasaki, N. Saito, T. Mibe, M. Abe, Nucl. Instr. Meth. Phys. Res. A **832**, 51 (2016)
[23] M. Firth, E. Habel, L. Krempasky, F. Schmeissner, Conf. Proc. C **730508**, 79 (1973)
[24] A. Yamamoto, Y. Makida, K. Tanaka, F. Krienen, B. Roberts, H. Brown, G. Bunce, G. Danby, M. G-

Perdekamp, H. Hseuh et al., Nuclear Instruments and Methods in Physics Research Section A: Accelerators, Spectrometers, Detectors and Associated Equipment **491**, 23 (2002)

[25] D. Barna, Phys. Rev. Accel. Beams **20**, 041002 (2017)

[26] D. Barna, M. Novák, K. Brunner, C. Petrone, M. Atanasov, J. Feuvrier, M. Pascal, IEEE Trans. Appl. Supercond. **29**, 4900108 (2018), 1809.04330

AUTHOR INDEX

EDP Sciences – Web of Conferences
17, Avenue du Hoggar
Parc d'Activité de Courtabouf
BP 112
F-91944 Les Ulis Cedex A
France

ISBN 978-1-7138-8045-5

9 781713 880455

CURRAN ASSOCIATES INC.
proceedings.com